A Small Book of Random Numbers

Volume One

James McNalley

ISBN: 1452818363
EAN13: 9781452818368

FAQ

Q: What is this?
A: This book contains a list of random numbers, organized into three columns of five digits.

Q: What is it for?
A: Creative people who would like to make careful use of chance in their work will find this list most useful. Composer John Cage used a similar list to determine the outcome of chance operations in his musical work.

Q: What isn't it for?
A: This book is not intended for secure cryptographic applications. It isn't a very good one time pad because anyone can purchase their own copy.

Q: How can I contact the author?
A: You can reach me at smallbookofrandomnumbers@gmail.com. I would be happy to hear about how you have used these numbers.

	A	B	C
01	46947	71735	94246
02	47417	72361	22495
03	63764	31439	69853
04	69586	04651	54047
05	64466	44369	54621
06	03030	85073	47591
07	97556	80617	38868
08	08858	18891	23055
09	58167	83419	52426
10	64238	97862	29802
11	68969	49254	93327
12	83410	76140	24855
13	65000	99048	91260
14	71572	87436	04552
15	91120	54017	26108
16	62834	51303	44974
17	19877	19006	52479
18	50145	21077	88350
19	35544	02278	50593
20	27762	63398	91656
21	90051	81339	79447
22	21231	94724	00886
23	19586	96535	94334
24	46988	11010	05888
25	06224	78024	69744
26	01536	08517	25475
27	77715	22620	96304
28	06538	42164	32246
29	03936	30843	94039
30	75374	44637	92084
31	79329	32383	08877
32	40739	38619	00119
33	02052	06432	30099
34	05014	14108	00191
35	06459	85787	45901
36	43265	45243	82906

	A	B	C
01	21981	82405	57682
02	73471	33788	18912
03	74382	45449	82215
04	25758	43699	64065
05	68580	05415	28425
06	72241	49581	44750
07	53393	20496	95931
08	25293	27050	05906
09	41084	36500	40473
10	51363	56957	12742
11	34312	47299	75652
12	86455	44642	47423
13	86833	24114	03359
14	86500	41215	38842
15	51098	73752	92522
16	90372	64114	01933
17	16957	35435	91874
18	47086	47324	07124
19	17940	48950	39586
20	64833	53937	92918
21	98920	03780	52054
22	43707	23350	29517
23	79771	35186	39166
24	27056	74117	33222
25	92618	42840	97592
26	45888	57320	16644
27	14311	02395	30731
28	51955	17148	15089
29	22718	05626	78527
30	55158	67358	38543
31	93516	30192	31814
32	32535	52393	81304
33	70845	25505	74838
34	31020	10036	53109
35	78885	58659	60880
36	30659	45028	76292

	A	B	C
01	18779	96085	68353
02	11100	69362	01732
03	98246	93219	32422
04	23933	57912	67565
05	88483	24073	34684
06	94004	99776	39526
07	96641	58155	02137
08	65674	88900	67343
09	30832	42412	06751
10	15904	16862	19444
11	89673	68491	40958
12	26936	57327	14935
13	61362	60022	15498
14	85418	62711	79482
15	57403	77505	56107
16	84640	14824	22528
17	83169	54123	92947
18	17901	44995	77183
19	11948	98427	91030
20	45637	35761	01326
21	53365	09816	75414
22	52152	95074	19651
23	51263	26969	25784
24	40892	55824	39174
25	75448	63872	27346
26	27130	49782	65300
27	80104	97728	77289
28	08312	57710	11644
29	87597	43090	84789
30	31520	16698	66299
31	17997	86717	94729
32	07171	89393	52526
33	61801	15010	05911
34	47751	55990	77329
35	54617	78597	21855
36	47219	54494	06697

	A	B	C
01	59325	89667	74158
02	74066	46716	82307
03	97875	70240	25848
04	20687	19250	33311
05	96515	59255	91307
06	80416	65238	36226
07	98095	89836	67981
08	93026	76661	63250
09	18186	52420	25890
10	75116	65520	30690
11	50639	28238	74185
12	71222	01848	15520
13	14069	92766	38824
14	40772	24819	71471
15	78093	85116	70374
16	97273	25713	63492
17	97656	46730	35271
18	87979	36162	89022
19	94047	31742	58353
20	35539	85585	35415
21	70575	77955	46220
22	66789	29882	66494
23	77475	54826	63571
24	71160	33648	12479
25	92454	20865	82754
26	02015	33972	96493
27	33182	68211	00738
28	13577	26624	46460
29	40830	67109	18751
30	18571	52250	27940
31	99053	79109	29886
32	36848	30843	50524
33	69164	11386	79447
34	71022	00543	99821
35	77615	81719	91844
36	09086	78078	57400

	A	B	C
01	84110	11605	94588
02	98450	81399	77337
03	17516	88632	17166
04	07481	12187	16078
05	64611	45959	91403
06	12963	72842	71216
07	79248	78980	20680
08	29453	43614	18205
09	44484	88009	22064
10	70502	85067	51074
11	19548	47883	09690
12	50866	17399	52706
13	59225	37308	70394
14	79922	99950	81126
15	47633	21366	37521
16	32731	85798	35846
17	55282	85965	17465
18	99395	95892	57994
19	27953	49130	26491
20	57014	89327	83143
21	57180	06916	96362
22	19104	55978	71928
23	07635	59132	54322
24	19017	73538	11149
25	07280	34106	11713
26	41921	83400	30666
27	97745	88480	26261
28	37393	45260	06452
29	86193	55051	44936
30	80883	57582	34958
31	62528	24131	15819
32	06318	88868	41152
33	22239	20499	59713
34	27045	12827	84943
35	03037	28455	34177
36	79664	89868	15570

	A	B	C
01	54704	97879	61174
02	23314	19834	20895
03	27789	56649	23129
04	82226	49518	13556
05	96416	04568	09553
06	20831	53250	49312
07	83450	21006	63252
08	83763	12413	43681
09	26879	08664	83926
10	69668	95510	33564
11	38186	06586	36863
12	69357	84861	81396
13	33492	47800	27501
14	34528	16047	78278
15	91218	30218	25744
16	26008	94478	21046
17	05129	57175	08175
18	86160	48177	81998
19	10723	55235	33068
20	23822	28307	76383
21	75021	81887	49487
22	99139	53467	78430
23	28313	79920	43804
24	67923	36389	77580
25	70840	51513	31894
26	79341	58700	95705
27	48325	12784	83196
28	28028	79066	29406
29	92050	53272	85369
30	16968	81048	20420
31	42058	17758	28414
32	72926	23316	23721
33	08353	35111	46996
34	60050	96538	53473
35	93829	95699	73055
36	28391	15921	46285

	A	B	C
01	08887	82063	54587
02	78215	16601	93371
03	07334	60170	76618
04	07034	37681	39083
05	32853	25627	27639
06	31120	98405	00259
07	59722	53050	47082
08	42915	07351	32101
09	42128	33446	52926
10	01134	18233	43949
11	45116	63811	23847
12	70395	33549	45911
13	08103	18712	88468
14	67155	81924	33711
15	97254	20208	18530
16	92520	57019	09006
17	55563	13690	16987
18	71863	36959	92985
19	91165	50085	18939
20	23321	98077	90791
21	94906	00353	65717
22	83671	07058	48898
23	17570	07663	87983
24	75214	76508	95407
25	18910	31975	80157
26	03050	22486	42749
27	83091	60908	29121
28	46382	12545	33368
29	47145	60263	66905
30	62402	15541	27039
31	26054	26077	81572
32	42808	71702	95421
33	68254	66218	93156
34	64269	15058	85489
35	51075	83037	38284
36	10433	22161	21887

	A	B	C
01	07263	80607	34322
02	51801	76183	92884
03	41902	47130	56104
04	44093	96620	71705
05	37308	69887	40225
06	06268	64825	61221
07	58421	39887	50229
08	40380	52727	29064
09	02151	76884	21872
10	93591	88671	35882
11	74906	19135	86669
12	29852	38302	71371
13	84038	65470	47585
14	54027	34107	47373
15	31908	45916	06291
16	43469	75593	55041
17	23316	93570	62304
18	68301	93700	92842
19	18631	55286	81447
20	99832	42916	41310
21	96874	22263	22382
22	36300	17687	68338
23	10560	27334	18081
24	46058	17070	07047
25	83089	70415	62562
26	93555	87382	21963
27	66238	76417	75799
28	41170	73860	70209
29	30534	26179	65631
30	97823	76159	16440
31	69006	47208	07700
32	87656	03315	46287
33	76731	00219	56550
34	21900	41331	20801
35	67413	01045	90263
36	97904	15306	36772

	A	B	C
01	24211	98050	23851
02	00213	01774	28948
03	61387	76392	85017
04	98550	51013	35777
05	32657	63874	00855
06	52890	17405	92534
07	34877	28908	04722
08	27088	18939	35410
09	99872	74485	09039
10	42859	37442	07310
11	80785	12010	53790
12	17929	10522	42065
13	34035	92423	97821
14	87639	76495	74844
15	91731	59171	76346
16	26450	41552	06399
17	45261	27035	30027
18	98135	11978	48011
19	29829	15142	18905
20	89857	09079	96037
21	23952	84129	47529
22	02842	49186	90144
23	21419	16245	40142
24	87041	80653	03660
25	56324	26697	96512
26	65478	59563	92351
27	37255	54080	38719
28	09293	64303	00181
29	52780	87359	49338
30	71133	68554	93729
31	34086	29455	59651
32	80053	73622	34265
33	97513	86222	72669
34	84720	28381	85492
35	05020	14972	61541
36	66096	31628	15984

	A	B	C
01	55187	37538	90731
02	72924	74302	96472
03	82671	46956	34835
04	22480	01500	54435
05	31292	59772	01086
06	73422	04995	27612
07	55113	27223	36691
08	29558	87610	59038
09	35542	67830	86046
10	50997	76610	86849
11	24489	01644	32264
12	21655	26785	23791
13	45418	22893	11823
14	10419	78739	27021
15	26066	70378	48780
16	19207	91551	45692
17	23818	02043	36847
18	82176	44257	01134
19	02982	25551	51675
20	06665	99223	83084
21	47207	82817	08012
22	33548	35337	83867
23	82392	83324	72387
24	55032	24330	77658
25	97822	54177	96711
26	04110	94877	48948
27	23434	77173	55815
28	32502	69948	74575
29	39032	03041	96947
30	24864	13655	27654
31	35431	14750	26853
32	47452	00752	16224
33	69736	01422	12809
34	79808	76611	39339
35	24630	09251	51010
36	95489	69262	76771

	A	B	C
01	09923	00551	61307
02	09775	47937	41451
03	11032	22297	79876
04	86678	09614	07066
05	89412	16177	42781
06	99265	90841	83840
07	84636	45970	76884
08	25920	06886	67387
09	40350	40434	64549
10	98465	00167	08378
11	28266	47340	07821
12	87885	20371	72020
13	22687	47789	73970
14	15657	10775	73852
15	08872	19020	37363
16	93139	49914	59936
17	44716	32483	79920
18	51637	36747	99453
19	86005	91350	60920
20	62743	58562	64954
21	02897	74815	95657
22	15555	89370	08883
23	06315	30365	92060
24	07958	14570	92105
25	05498	52223	42736
26	81731	45518	11779
27	03417	69687	86274
28	09398	69641	56952
29	71278	86761	82936
30	80137	54265	19949
31	89927	33649	24954
32	10561	76101	62479
33	79068	80735	46848
34	84651	39089	19982
35	22486	30422	24319
36	43678	80042	30243

	A	B	C
01	08493	70712	95431
02	48709	72923	88539
03	41353	69349	22387
04	63516	33761	59832
05	20095	42983	03649
06	58154	45835	14132
07	16102	99992	30500
08	36651	64893	71152
09	33066	13255	15396
10	27556	17256	57187
11	80698	34536	79574
12	78017	57902	51563
13	36004	10906	72325
14	24235	90642	85216
15	39224	72878	91681
16	33365	67352	84390
17	91861	74974	64308
18	26145	00539	75259
19	58564	37942	63472
20	66426	41828	18507
21	40650	81628	17047
22	67543	66548	40299
23	33211	75033	78570
24	17033	80548	13566
25	95516	91727	37276
26	27111	30592	23660
27	91830	36431	60623
28	01183	34706	04015
29	25623	23724	34851
30	43705	93173	42009
31	90393	99971	26625
32	98366	37364	29885
33	24089	46732	42317
34	68764	47508	98607
35	96633	81908	53578
36	39569	34305	87574

	A	B	C
01	89694	52599	53765
02	86252	51341	00259
03	19499	39578	26485
04	82911	98507	82033
05	06401	12533	87153
06	85260	73474	32966
07	89181	75384	14851
08	63090	62375	51518
09	87263	41456	55486
10	43712	45002	50079
11	98059	49398	74006
12	44258	75990	56148
13	07279	79582	35567
14	14558	36454	17859
15	26688	28636	40015
16	82216	49096	21221
17	14726	50201	17520
18	53060	95740	85014
19	31391	84142	14471
20	88509	17923	93750
21	96636	46051	87038
22	36013	59969	90272
23	01834	12356	33277
24	20988	00837	84729
25	49897	33145	02953
26	15475	81839	92924
27	97075	04951	23182
28	63422	38417	00790
29	18149	57513	80603
30	29927	32997	46414
31	81685	91483	73663
32	03892	98080	99515
33	65116	52622	18619
34	57354	80380	06535
35	52085	43716	26122
36	66413	81886	25826

	A	B	C
01	21995	54083	52785
02	38847	59716	42994
03	87027	89044	97141
04	90330	70353	78480
05	79684	41704	45160
06	38103	24637	85020
07	72035	49074	42546
08	05511	71196	72250
09	60435	93591	91853
10	20830	61597	62742
11	69813	94052	21702
12	87397	94338	04771
13	38667	85804	66549
14	02435	46700	89091
15	08283	28941	04358
16	83192	78185	67744
17	53589	50856	92311
18	25527	49912	91949
19	13571	59797	08048
20	87713	27310	56574
21	22528	50178	38933
22	96311	18704	84877
23	69594	53448	70859
24	40653	20432	82039
25	64926	96103	03082
26	92319	69086	16761
27	30986	77283	87874
28	44383	46563	04829
29	70400	83540	00736
30	85265	16845	63478
31	75989	96071	30469
32	38625	33853	29063
33	44527	93476	29102
34	36925	07464	19706
35	41657	94167	57102
36	08540	71961	53876

	A	B	C
01	93499	84378	78092
02	54401	77386	65542
03	62732	09588	05536
04	08636	22072	70145
05	31690	90844	17907
06	50166	85131	47843
07	90221	90709	44027
08	13072	22063	57016
09	74619	31193	32722
10	21748	55573	63385
11	25225	90312	44049
12	23607	02095	45173
13	36997	76055	99397
14	21377	96248	27255
15	20377	10514	87259
16	80474	08657	48375
17	45264	16296	13174
18	83720	91974	87983
19	94957	49324	01424
20	24736	43803	16380
21	53186	10512	79340
22	38097	79377	88773
23	52457	75683	91527
24	42428	17395	30560
25	18949	69704	12130
26	43598	00982	60073
27	66123	62780	29805
28	35664	50594	95884
29	81810	44355	38976
30	59164	22503	76944
31	37069	99238	75634
32	53191	54590	24434
33	73927	36117	84588
34	97014	85706	30934
35	12045	26109	05979
36	09633	47832	22632

	A	B	C
01	64168	88020	20990
02	40807	88818	51435
03	72345	79391	70052
04	59506	58727	20293
05	47460	48905	27195
06	92478	66026	35553
07	98124	10620	44697
08	12089	38375	66129
09	21845	54505	32841
10	90245	55075	04387
11	98312	37760	52034
12	26161	25882	69520
13	23036	86999	82994
14	79994	68379	22524
15	44763	57980	33920
16	54618	72660	61317
17	63358	53533	76039
18	83069	11913	00097
19	37422	97338	75936
20	14584	28066	33851
21	50332	82985	40606
22	46063	90242	86729
23	43983	41573	96910
24	49739	63167	94275
25	90329	18326	07875
26	57695	82534	02313
27	75712	72881	05248
28	72198	80514	63598
29	34142	23788	25208
30	26397	87455	10989
31	10068	73669	61859
32	53437	44579	88622
33	72198	26936	28601
34	20141	12552	09680
35	05405	94213	60680
36	15375	77018	36380

	A	B	C
01	29789	43877	95984
02	87878	59301	90245
03	43729	29356	39390
04	52432	33047	01645
05	13722	96625	10050
06	71295	87799	41566
07	18982	62413	82790
08	95647	81345	47435
09	51331	01074	30542
10	15002	11718	70792
11	02007	54156	32298
12	62356	90084	41992
13	36406	01250	60547
14	90195	40581	15443
15	82649	23396	86863
16	95240	00944	75556
17	62427	12922	75177
18	67307	38786	24273
19	47504	97327	37421
20	39531	32244	23554
21	81326	89419	74424
22	43393	78547	81965
23	09562	99481	30728
24	77580	61805	53360
25	80438	32432	35526
26	76288	28547	13971
27	95284	77498	40078
28	98049	11526	51059
29	82975	46571	50153
30	56595	38508	55250
31	74088	50574	20921
32	57691	70249	65319
33	10847	18274	68566
34	81980	52220	62463
35	38625	56725	64095
36	80422	05791	26734

	A	B	C
01	60015	48062	65600
02	16465	48502	59532
03	45896	14060	13240
04	84459	19407	19312
05	31724	76478	89386
06	92939	45255	72427
07	21587	24766	29756
08	47453	69620	27507
09	31577	13697	99928
10	17749	82804	72594
11	46745	24910	99812
12	79168	24527	18142
13	28460	13730	86553
14	90252	77980	05677
15	13554	32462	70518
16	65290	51225	38193
17	13294	36154	15198
18	74528	03064	16590
19	01982	32214	51707
20	11037	03648	54424
21	70305	63343	16001
22	73760	94876	10183
23	08921	09098	44859
24	73984	42535	40941
25	16770	64545	79167
26	16867	22945	78368
27	92593	24733	45451
28	60514	64377	28362
29	18046	19849	45573
30	75684	48596	62746
31	60643	37501	26927
32	08136	48472	08221
33	28682	90522	99178
34	73718	84050	10184
35	14783	98766	00626
36	05511	76750	47238

	A	B	C
01	92960	38813	45098
02	53835	37293	06277
03	83850	10369	59728
04	88421	34980	74165
05	50454	88202	42205
06	73273	47110	59521
07	30391	94974	46280
08	94543	01811	33016
09	45742	66933	00453
10	40696	68016	88493
11	12054	52482	55797
12	07813	30349	59522
13	48126	09092	61007
14	83155	42123	70323
15	34150	96440	50859
16	55470	18678	35997
17	29008	83615	02021
18	97718	32339	40721
19	16092	65828	73235
20	61254	87178	94559
21	30605	77313	57007
22	62974	44444	78862
23	33816	40361	84385
24	28206	90526	78152
25	34938	42506	15865
26	38739	49173	30901
27	62921	20758	14845
28	28951	19509	12020
29	00540	88429	61063
30	29986	29913	14250
31	59484	40766	20655
32	79486	38006	90516
33	40086	69718	87662
34	81389	03682	92293
35	94260	45655	81496
36	24649	21316	77619

	A	B	C
01	55261	51760	32916
02	97670	16407	32607
03	21172	92197	16964
04	15375	76617	68905
05	15153	59111	64449
06	63092	97480	22819
07	67453	33500	26340
08	07069	27092	11577
09	97522	14535	73103
10	45318	31925	43284
11	46338	15575	63700
12	29604	92687	89306
13	64633	66943	01816
14	79939	03186	85021
15	72139	81386	03434
16	59789	30213	49965
17	77891	91023	75524
18	13623	44603	98224
19	16237	27012	06476
20	13456	80384	98396
21	63146	74317	48592
22	34121	46675	62175
23	26222	64225	85039
24	75209	73442	97655
25	21599	54080	69300
26	66248	22403	74331
27	05853	25627	69600
28	84695	86724	04277
29	37539	00894	55525
30	60254	11608	93219
31	32706	31426	14231
32	76277	81879	30351
33	01326	53444	37954
34	54998	66576	72650
35	82684	04994	98133
36	90710	31538	01308

	A	B	C
01	11912	56019	36163
02	36803	67123	98207
03	02980	31146	78386
04	74282	82782	13522
05	74888	69891	13405
06	35645	86592	12712
07	21767	21885	86742
08	76795	54065	47371
09	75441	99666	46284
10	33462	81079	05233
11	85099	97326	66004
12	37310	07105	12694
13	79637	50706	54192
14	66867	13760	06978
15	45110	11598	20487
16	90097	03713	16495
17	75542	73020	02862
18	36127	03110	31675
19	00985	49360	02386
20	18026	08158	39506
21	94494	74390	12817
22	90043	17495	30920
23	68836	86283	02376
24	33564	89010	29217
25	72479	86200	92811
26	31651	35658	27200
27	23555	00356	52769
28	14810	29696	74806
29	06505	06215	03398
30	12876	11650	29209
31	72430	01147	48739
32	98117	50347	72379
33	62701	64250	97650
34	30732	57961	43176
35	56333	62948	95937
36	59274	40034	40002

	A	B	C
01	65334	78895	55803
02	69497	55129	53383
03	93946	95666	74518
04	81893	73638	84442
05	12955	08103	75220
06	26964	74827	88918
07	17311	89511	99428
08	38743	55135	22195
09	42481	35116	62532
10	39903	67979	79127
11	32442	07855	17018
12	19051	35741	24301
13	76378	22346	07724
14	95807	05411	26542
15	08057	46529	83794
16	46450	57277	74513
17	03777	79139	47827
18	73207	88437	99501
19	27641	56563	32874
20	06196	54364	13781
21	63090	54644	44983
22	18920	39591	96751
23	78924	04605	74116
24	45163	71657	75554
25	08578	89280	14750
26	39270	99415	34898
27	81785	08445	11540
28	02859	64821	86562
29	43973	76374	41529
30	56775	45386	70438
31	72219	14046	38245
32	94245	40624	99980
33	14210	75746	01852
34	69195	17044	14052
35	05881	11401	34633
36	60617	47890	13378

	A	B	C
01	22516	91865	08321
02	28557	39631	69436
03	40814	97742	54160
04	93725	78716	80989
05	04697	55805	97222
06	01652	32238	55608
07	74886	42366	23286
08	49625	48904	57513
09	05440	89632	79945
10	55166	97759	05587
11	96888	31852	19386
12	09129	16673	20119
13	50582	57210	63982
14	50176	52807	22340
15	65635	24882	99900
16	46522	45721	12792
17	56961	58721	44498
18	05195	13506	38770
19	39368	28914	63518
20	58964	81214	80204
21	23558	20108	12222
22	96879	25557	53393
23	38925	97043	21621
24	93363	46528	88823
25	50528	31088	63948
26	20625	47789	85708
27	26627	34727	55545
28	24121	80586	10635
29	09243	08923	89383
30	64456	71759	43572
31	78505	08965	58975
32	56143	01356	92018
33	95693	76732	69474
34	82977	29333	12835
35	10317	92883	85860
36	86539	60007	01787

	A	B	C
01	35572	88642	83118
02	45697	55788	73720
03	63697	08510	99024
04	05560	32068	19555
05	92162	25458	05872
06	09148	59365	60100
07	14723	89731	17071
08	91562	29613	29739
09	42717	10308	98871
10	30313	05042	47479
11	48299	59614	78361
12	82776	45673	17890
13	05528	07341	49105
14	40227	91957	68364
15	37127	07428	25052
16	08248	63550	06191
17	99007	22844	57437
18	21475	47604	01211
19	78363	74701	02809
20	82622	41601	41903
21	81611	56536	90256
22	34983	17692	11310
23	24216	42460	50359
24	06820	66317	66440
25	75710	38286	50209
26	65279	98588	23547
27	71574	00599	66910
28	67746	42786	31901
29	95698	32446	73697
30	36435	45901	61061
31	48027	16027	99698
32	83266	04402	08534
33	54533	01420	35912
34	19147	26947	38791
35	64381	24360	99048
36	27086	29787	52962

	A	B	C
01	20769	21524	34518
02	78571	74731	66849
03	83707	11337	51166
04	88806	18362	18009
05	15402	44010	13664
06	00531	93468	59412
07	42887	23893	37107
08	36320	56023	88960
09	72277	19406	97369
10	08117	16497	94594
11	17003	79404	42540
12	02615	30435	02801
13	59270	97899	74888
14	86877	10983	73374
15	60193	23232	20633
16	59912	08864	31172
17	85168	18442	04016
18	72490	46938	16134
19	03495	20832	00152
20	35914	20103	62767
21	48539	20075	73980
22	55266	41680	33524
23	37405	47466	25520
24	71877	62028	84966
25	48215	77774	17813
26	96795	63692	86006
27	91141	77773	63781
28	63343	15821	10414
29	07991	68565	87919
30	61745	62122	83350
31	48375	55071	82260
32	42870	63597	35623
33	83610	05425	50841
34	62303	30247	86983
35	45690	40410	79459
36	39701	11594	10952

	A	B	C
01	48945	82427	49434
02	75904	51324	70351
03	00323	22289	02327
04	74358	50912	75407
05	01216	51560	87751
06	50963	55877	49743
07	81923	11062	85280
08	32498	24761	32178
09	63686	70109	59342
10	63701	91703	72460
11	73872	00337	83171
12	53612	72532	58671
13	66143	08203	19473
14	22869	65817	71924
15	21502	70434	59588
16	97699	41920	09537
17	67374	91353	43657
18	13391	87987	79165
19	21551	50861	85844
20	45425	42203	46038
21	90952	06629	65347
22	45162	04274	80279
23	87710	13995	01568
24	81727	51896	40643
25	89671	58860	81298
26	70118	89406	51199
27	44909	94135	63915
28	40834	86294	92251
29	44539	96358	49118
30	79744	52355	29796
31	00258	00850	54353
32	63371	46905	26772
33	96578	77928	18915
34	37238	07170	52870
35	78740	82891	39885
36	54126	43620	94196

	A	B	C
01	24179	75992	05204
02	66979	84722	50346
03	45604	08994	59080
04	42392	82533	12401
05	75144	54656	93529
06	14508	34475	53648
07	07828	06873	53185
08	81555	90001	04812
09	03116	51181	51219
10	23202	43542	29190
11	56904	54279	87849
12	72917	68276	13450
13	56372	28317	74788
14	63886	25658	00378
15	61864	91470	19517
16	16087	78214	41853
17	39904	98181	23870
18	38742	21300	19868
19	36332	62413	46356
20	45638	53006	94405
21	73074	13415	99198
22	74682	17131	80296
23	07764	32887	33318
24	17687	94189	39764
25	44357	93876	27902
26	02744	72274	85115
27	36770	82441	51564
28	98557	84890	09600
29	11981	91895	61728
30	36411	74618	32260
31	06863	60943	00381
32	01808	83732	99468
33	07317	68551	65731
34	27251	81112	90783
35	99672	46071	43174
36	37431	76346	89699

	A	B	C
01	74339	63292	74072
02	58672	94834	92274
03	04931	16241	14359
04	41220	12665	40227
05	98523	29605	45464
06	67998	13353	37809
07	15283	11413	46621
08	56022	74251	89678
09	20296	40347	49756
10	41962	36577	54405
11	77651	69595	55641
12	00559	84750	34714
13	29450	34299	15329
14	24774	70186	53149
15	35075	18500	28876
16	37478	29140	47635
17	71203	92302	82431
18	68217	67774	76296
19	41515	07465	45436
20	75899	62857	17035
21	43486	62428	53106
22	13895	39416	42315
23	33537	00113	11927
24	50916	17827	98855
25	28329	15329	64043
26	04279	07780	07614
27	01118	03777	47787
28	61334	70790	42112
29	65147	48541	12680
30	62424	36372	18341
31	43495	83219	72179
32	01964	89688	27290
33	84664	65072	71188
34	24724	50525	46687
35	51834	62135	65774
36	08438	55652	08588

	A	B	C
01	36631	83772	51859
02	88583	42510	73462
03	07572	32399	92140
04	20393	97039	97538
05	04610	56529	51006
06	49912	57218	40286
07	60991	31870	96416
08	09995	58250	19161
09	67121	13759	70882
10	49324	84592	69807
11	92124	81929	87857
12	48247	77349	10514
13	29846	54081	03435
14	01799	43339	26490
15	03823	13563	95199
16	67268	85738	34372
17	09192	16153	80771
18	29433	56781	16731
19	14885	71677	84707
20	87984	81805	30521
21	02538	79846	87984
22	36981	64642	58065
23	77316	11813	58708
24	53170	89994	57916
25	56433	36900	51695
26	28764	30386	59485
27	24831	27126	11133
28	35429	41586	40953
29	74941	17990	59126
30	71246	56083	05542
31	71252	17499	06717
32	70144	15073	71583
33	84766	75341	00499
34	89275	89830	39889
35	12238	16946	12963
36	85782	14877	04146

	A	B	C
01	52055	41483	97740
02	78244	31135	34378
03	42863	38445	91369
04	62470	69109	78422
05	05120	26713	50884
06	94825	58673	73862
07	01742	66703	03346
08	05160	72741	34253
09	83222	91278	72534
10	20277	58123	15039
11	34340	21504	65955
12	25337	78032	65522
13	91992	13572	04850
14	39633	97348	76705
15	44135	41442	15608
16	94716	42284	20511
17	90949	62569	25936
18	64132	00338	44569
19	82369	57499	31537
20	62150	53753	18033
21	20569	94609	43045
22	75582	66233	86919
23	50701	24175	90589
24	36186	69839	56591
25	28356	61546	30340
26	62701	41333	54446
27	20893	19246	27814
28	86100	25078	21826
29	12989	90794	04315
30	18421	59476	46879
31	84581	13793	31031
32	92067	36800	27968
33	02589	57524	51249
34	80026	32672	82213
35	13502	15690	70567
36	32004	19025	43011

	A	B	C
01	76507	68987	02091
02	22509	40409	80417
03	48223	90062	02262
04	67103	92443	57794
05	98016	74112	68295
06	09396	44016	25381
07	06187	65742	98878
08	59780	14932	51105
09	01648	99559	51044
10	84640	40642	32798
11	26000	02556	00658
12	34360	24385	05990
13	68191	44698	38547
14	10410	21096	83353
15	49302	54001	12860
16	60872	78131	54172
17	56449	53449	91143
18	30547	34252	02328
19	90570	60638	72657
20	85514	57607	06331
21	29081	34634	20499
22	00166	49801	43368
23	82488	47870	96981
24	91854	81731	66748
25	64055	79334	30817
26	54924	72832	42629
27	87256	86489	74446
28	93069	78730	83863
29	39976	75700	13918
30	43208	26525	35482
31	32664	70992	99773
32	65694	14010	93313
33	85692	69206	17880
34	31218	25449	03833
35	46216	54362	12704
36	62864	96395	42161

	A	B	C
01	63660	31804	31142
02	36171	82063	28282
03	78408	23221	53285
04	30488	19451	55134
05	39844	55238	31701
06	42456	69732	28118
07	44957	43882	63098
08	49820	18231	61576
09	93956	56090	67515
10	81438	63906	61163
11	18239	09210	65381
12	81677	85625	91005
13	74344	87698	93407
14	65028	53568	57458
15	83323	11487	29001
16	77453	77176	07681
17	81696	28725	80164
18	90419	64748	35734
19	72172	82035	11102
20	20799	62222	03161
21	19126	10362	24921
22	80062	06193	75508
23	73927	53247	53301
24	62258	17352	51763
25	97260	80389	36629
26	47936	36046	29592
27	19755	64164	74557
28	64229	91460	95356
29	51908	54406	02424
30	50234	72326	98471
31	58358	15124	23071
32	42138	90584	52602
33	70063	03680	96516
34	63309	16808	89169
35	00235	88517	13523
36	89102	39754	04353

	A	B	C
01	89284	79445	85071
02	07520	84673	23504
03	97074	10988	31946
04	65012	90560	09763
05	60300	31853	30910
06	09255	93689	63254
07	56304	30899	45575
08	94596	48541	35874
09	57779	54423	97851
10	63646	14420	50968
11	27642	80221	94525
12	62164	07501	65397
13	78921	36722	76768
14	66922	05830	41074
15	26899	58527	52975
16	26760	25547	12988
17	54039	45384	84228
18	10930	96242	73909
19	86847	45183	54953
20	34362	17070	09627
21	93463	05397	00238
22	47077	90743	50534
23	26116	72231	41962
24	30112	32258	35894
25	23791	02948	37392
26	44211	57195	75672
27	18270	07668	05611
28	84837	41615	34559
29	80921	21399	13632
30	81358	35436	66119
31	51275	05986	13028
32	28254	70276	23630
33	06022	84891	29796
34	28126	82239	05564
35	88489	54068	56886
36	51900	29693	60624

	A	B	C
01	30697	80926	02951
02	11521	17658	07037
03	13779	61002	04268
04	77656	67812	32970
05	25626	88295	76889
06	33274	68523	70378
07	08389	61269	14322
08	30102	33895	51953
09	39251	74594	15953
10	40265	44280	03516
11	70904	10059	21153
12	07946	70187	02117
13	68604	16462	04776
14	83635	02979	33670
15	91051	35780	29225
16	16684	17912	02009
17	05205	81368	35789
18	16516	96007	41941
19	34046	99641	23827
20	75763	85992	02195
21	95605	64738	38367
22	64893	73349	80517
23	06999	11743	68666
24	28393	45877	12061
25	18433	84917	97861
26	25325	37911	12592
27	84474	51673	15583
28	42253	97336	81239
29	44252	54454	50899
30	80303	16777	23339
31	00016	51874	58101
32	80539	96787	04296
33	66498	58633	38829
34	19728	91719	53067
35	37725	75639	61348
36	75334	37992	35567

	A	B	C
01	24123	46750	29988
02	47769	39759	62595
03	14454	42183	94875
04	22931	23482	59294
05	81781	53694	34021
06	42167	30649	67757
07	85273	41566	78537
08	64570	48903	04217
09	08691	62236	86172
10	47078	80768	35593
11	14057	09597	47871
12	19915	75609	21973
13	95953	67688	39169
14	19027	64805	58255
15	85640	78827	48105
16	37989	41539	63358
17	62582	97315	79099
18	34772	67557	17049
19	37273	80339	73260
20	14804	09307	28146
21	91608	27400	20342
22	67384	67999	87474
23	30258	18659	97691
24	32471	32701	05520
25	62379	29731	57401
26	81166	28664	30370
27	71922	54144	57254
28	30959	51071	11392
29	64273	60763	49939
30	98688	20838	48494
31	74877	71986	63949
32	98192	18474	33062
33	68647	52597	46260
34	90120	11879	51385
35	82386	71513	59459
36	53283	86945	06470

	A	B	C
01	30989	52419	54372
02	55365	96417	75478
03	29964	89683	26149
04	35793	55701	12869
05	82539	60882	74205
06	76699	77275	08384
07	72902	60682	19465
08	68424	30835	41831
09	70471	47582	48594
10	82552	04556	20700
11	52316	70542	24319
12	74037	39317	36237
13	20507	88895	30231
14	01099	54163	56266
15	24507	06326	23466
16	46764	32496	49487
17	41711	78386	08754
18	65467	54433	18561
19	01969	63714	14756
20	71096	23846	04706
21	95258	85544	19705
22	58900	96158	34026
23	38104	20809	92890
24	20474	22366	99372
25	93917	97697	00569
26	25207	78671	10270
27	15535	76610	60475
28	12343	01188	60055
29	45484	65717	71730
30	20099	88803	36073
31	59883	38373	25820
32	33318	99282	80859
33	89258	61137	47875
34	68214	96298	65548
35	51617	17136	97031
36	95912	42007	15685

	A	B	C
01	09124	56095	21045
02	37943	14543	25877
03	45014	18673	06147
04	43079	54279	99352
05	84541	94265	08862
06	51410	71083	13996
07	11358	22556	33541
08	37088	51807	20183
09	60132	92517	42944
10	53858	30668	78650
11	55449	04403	40290
12	51400	10022	85245
13	13162	81185	50180
14	58883	73775	14804
15	13596	62952	21725
16	14122	90009	09740
17	71761	05469	03871
18	88104	07126	38991
19	21733	57173	49433
20	83091	12798	76208
21	88067	33527	23630
22	83291	07369	28097
23	93085	86156	35512
24	86890	15428	37260
25	12418	04304	44597
26	98960	02656	65817
27	48976	02641	14843
28	74015	39747	33743
29	40642	38491	65582
30	56958	24853	31003
31	50955	68764	06294
32	75343	90112	42486
33	34845	20039	00983
34	23702	92134	15276
35	47325	83619	67296
36	05580	87471	93322

	A	B	C
01	12404	83685	34274
02	06582	28215	02462
03	43940	01228	91960
04	26738	08215	10745
05	47178	92211	04318
06	06774	52716	28156
07	68620	10833	66880
08	84923	80303	75864
09	35106	14371	96091
10	02442	58340	20618
11	17287	40448	16894
12	37997	22977	83092
13	96576	73550	27844
14	93284	03231	77636
15	02182	78437	84600
16	80598	39841	71365
17	57995	59911	34968
18	16914	00231	62304
19	07125	79727	89263
20	92572	70828	03066
21	99874	62851	57870
22	35630	43799	44680
23	30212	03031	00895
24	31380	94878	63429
25	32561	33136	67487
26	62902	69490	40701
27	62503	05657	90364
28	16814	78354	90728
29	90990	11802	76573
30	75844	26682	74208
31	25615	28493	28718
32	23887	01167	29544
33	07074	73358	24850
34	56524	66546	04425
35	11196	07172	20417
36	24439	07921	48429

	A	B	C
01	12009	25694	90147
02	20285	70307	87535
03	07256	32836	90104
04	38946	16273	13014
05	34634	33311	93659
06	57683	94567	32094
07	23355	29475	59975
08	53890	46282	30832
09	94968	92073	29854
10	25997	64335	38813
11	63623	65316	82071
12	75841	73738	76533
13	31891	39212	04971
14	67238	61314	46619
15	15118	62371	35008
16	74527	94983	95210
17	51679	66944	94343
18	23033	46467	14377
19	05277	27025	56120
20	93511	95745	10805
21	39677	51557	83748
22	60679	19105	22291
23	25273	58890	54695
24	47898	38220	71668
25	28532	36841	18018
26	45776	84982	41089
27	68855	23504	55813
28	48897	62387	71771
29	06437	56925	59513
30	33144	68129	52009
31	84953	45744	92988
32	45910	11560	94776
33	25203	48268	34481
34	43879	00518	90772
35	19577	39084	05184
36	79122	68951	27305

	A	B	C
01	21914	12276	02488
02	06769	43285	45095
03	35536	53842	26171
04	35691	10143	65496
05	78748	55994	02649
06	20121	94889	49279
07	74891	78362	47051
08	15485	55478	12642
09	16334	89980	64236
10	70976	62557	23895
11	13259	27864	79678
12	86955	81626	54163
13	48153	22068	81392
14	61023	06678	21998
15	21578	41332	33780
16	94151	28042	41244
17	27500	22227	98390
18	11723	85359	67816
19	70686	62706	40229
20	54896	88808	63952
21	29141	26504	39072
22	08892	91501	68345
23	34822	40215	77684
24	28695	45621	86198
25	01141	12812	71638
26	72293	80565	21773
27	24802	20357	73422
28	07469	48788	81385
29	05471	22025	99011
30	11449	04544	41386
31	47660	96565	18902
32	67519	29922	40887
33	69844	67994	84822
34	86491	83367	17217
35	12011	53873	07364
36	41811	48153	20586

	A	B	C
01	21883	65071	43023
02	68885	27286	54195
03	68667	49862	71481
04	31351	19916	79611
05	20983	16956	95948
06	64182	92968	82211
07	18781	04653	72542
08	90430	50722	29153
09	73295	61743	53799
10	28389	83895	18913
11	76310	60799	72974
12	93926	69127	10796
13	07519	70643	53274
14	89558	59643	66941
15	14884	87463	61223
16	71842	43282	53579
17	93974	13510	05110
18	62545	23393	01339
19	02399	69993	97438
20	61490	11659	87332
21	20702	54660	38489
22	88136	56483	08252
23	30104	59677	34149
24	11345	22991	29988
25	92836	56688	12602
26	08650	80680	26484
27	06421	41836	25058
28	73830	82334	57305
29	55073	58450	06158
30	74916	93357	63855
31	01822	83027	01843
32	94098	33943	04440
33	60313	68830	92491
34	64958	70909	72115
35	97757	71781	59584
36	73858	18149	21863

	A	B	C
01	52661	49248	49334
02	92097	39783	18247
03	57837	48497	39507
04	79106	39940	62380
05	29901	63374	83265
06	29400	17653	37910
07	88164	41653	15349
08	31115	78734	34639
09	51968	19185	07631
10	46785	92018	75069
11	71900	40739	38977
12	00212	78800	60010
13	42440	68759	52210
14	95309	26143	77251
15	71325	98894	90452
16	87202	82800	20311
17	64380	59364	53511
18	86703	20181	45387
19	30814	50967	34785
20	77853	01301	92622
21	29747	54303	24339
22	95462	06357	03954
23	14604	41997	61647
24	14317	38613	82184
25	12711	45863	70163
26	83326	58231	08222
27	97938	33192	15950
28	27963	20214	58234
29	04881	20490	74359
30	50307	36495	72505
31	66744	03736	65623
32	92242	89540	18225
33	83362	90035	29205
34	25999	01122	97641
35	26587	57823	58092
36	27601	75876	12194

	A	B	C
01	33543	97519	25925
02	38927	52919	47290
03	15168	58073	05371
04	40042	92822	50091
05	55427	78356	62355
06	59363	51401	96956
07	02530	90265	03485
08	97487	19654	63592
09	96052	65926	85618
10	55840	20664	66088
11	74893	11823	96439
12	87444	64395	65398
13	24480	74200	87520
14	77977	23257	34262
15	52020	96136	43079
16	42941	78892	52060
17	48978	01224	58833
18	07279	82198	37627
19	79927	97765	03430
20	74735	44754	79408
21	75597	41209	56761
22	56431	95977	45148
23	02132	31731	14043
24	74786	94252	36406
25	86174	14815	25700
26	39580	83330	35803
27	29332	65507	74930
28	78301	62403	73602
29	62514	77364	85858
30	60815	55400	48755
31	94172	05704	43341
32	28543	95885	71863
33	39833	68451	71489
34	52578	29921	34280
35	98207	56186	40606
36	55638	62232	21947

	A	B	C
01	43473	54267	86840
02	49473	31041	35277
03	27580	68940	89376
04	23607	35782	28517
05	74930	08111	99368
06	83423	68116	09410
07	35407	83224	71975
08	91005	70601	24735
09	26532	22379	13381
10	27513	28018	78069
11	65049	86690	89261
12	20182	51218	15857
13	85871	44454	78088
14	19963	65402	65520
15	00416	54863	27393
16	74137	48287	18852
17	46361	39372	31843
18	33582	77205	51289
19	45430	77731	99899
20	25651	13848	33225
21	82106	19592	65546
22	62314	91448	31443
23	69579	74943	31205
24	60520	69677	37938
25	30100	05407	47434
26	17566	64019	68010
27	74569	34155	28694
28	41355	25628	58056
29	97257	68515	95949
30	17959	31008	31765
31	16804	51656	45568
32	10827	52174	30671
33	51805	59991	19679
34	46220	97933	69649
35	97386	79777	50986
36	33430	31440	25832

	A	B	C
01	58482	95063	48350
02	20221	30656	23345
03	54506	06571	23024
04	12477	19358	60110
05	48859	16069	88050
06	40113	58295	83740
07	95766	79222	65938
08	93405	32846	44103
09	23375	76251	16500
10	49481	25830	08787
11	69622	50824	42571
12	97455	53171	31946
13	20193	80719	29880
14	14238	47073	11170
15	44383	24926	45847
16	06187	30885	69196
17	89109	32097	48935
18	43572	83510	37438
19	73937	75902	94854
20	89949	26835	95959
21	07281	47056	18125
22	30464	39826	46740
23	51778	33748	19293
24	32137	58656	59749
25	66176	08139	32463
26	14735	86797	41283
27	99545	76729	36349
28	67500	98517	03527
29	24302	34577	11516
30	18380	62237	27782
31	88552	39083	58400
32	51395	69380	74688
33	83632	92904	13837
34	33763	38071	48775
35	18896	10082	45736
36	01864	35160	72873

	A	B	C
01	43885	41103	61304
02	43181	21468	67863
03	96355	96499	61934
04	41389	48416	14371
05	99339	55328	91955
06	00479	91717	01066
07	60407	78992	30089
08	29773	41249	17865
09	99417	22713	91266
10	37475	71497	52595
11	15582	80556	89399
12	48937	43171	55986
13	17183	45211	91029
14	02421	07564	26269
15	22558	94900	74321
16	79724	90472	51475
17	61150	37482	97540
18	51527	58360	60454
19	47939	17668	23049
20	18925	48134	11805
21	47316	00039	23800
22	27416	91832	13420
23	42424	15557	97780
24	81293	77371	26446
25	12945	27783	23071
26	29205	12403	45233
27	61136	34310	79910
28	63419	74061	36411
29	22125	85194	75760
30	07467	79087	87448
31	23383	54785	85231
32	95957	95857	62884
33	31871	21811	21560
34	53947	09158	28086
35	13082	42906	25620
36	09338	49831	71924

	A	B	C
01	26277	84075	82428
02	95410	33299	88312
03	96074	68367	64019
04	90160	31455	86193
05	65799	51178	61046
06	90596	43293	95277
07	27932	15338	02756
08	48073	42804	95520
09	19091	51212	27305
10	13835	56655	87030
11	54846	26539	12341
12	77725	99977	37300
13	16387	62833	66436
14	19522	45615	52350
15	41924	68549	98498
16	09698	94177	56807
17	49488	75169	97402
18	69363	48361	80061
19	30136	67567	99530
20	17643	38862	71792
21	37409	17344	27982
22	74554	99647	68514
23	17619	95578	31883
24	30241	97579	80670
25	54594	80458	70104
26	03042	20777	31316
27	02242	16463	77938
28	40876	29116	11900
29	80478	13053	71349
30	00460	56908	14232
31	31332	40267	24746
32	80018	22720	59069
33	60865	34113	22817
34	35287	98374	98780
35	75811	97886	04805
36	98664	31342	18494

	A	B	C
01	83225	12936	78025
02	46414	07404	04142
03	29721	95094	38743
04	00921	50514	16615
05	63116	33753	52170
06	16247	58510	03252
07	31886	07124	55318
08	81750	94331	16922
09	48184	55413	98596
10	53482	86711	17496
11	95672	58923	72681
12	04660	39999	82168
13	96463	65710	40212
14	06147	57356	81890
15	20896	77969	91464
16	70407	65060	87172
17	17863	00391	39607
18	46938	75778	98813
19	90036	75785	82282
20	48520	99143	76387
21	66453	29182	09267
22	94145	41034	46406
23	31291	26268	11549
24	06515	08711	18933
25	34008	54048	93500
26	81484	35635	22489
27	43870	53804	20493
28	85004	07823	34535
29	98325	21350	33023
30	88475	12990	99556
31	35957	13909	93725
32	72507	33229	14911
33	30500	72060	88303
34	29607	85234	28842
35	97979	61464	48188
36	80027	55615	01008

	A	B	C
01	76479	36482	61627
02	23621	59283	35626
03	24346	74622	30612
04	05928	71963	21481
05	53920	23716	46408
06	74039	02720	01587
07	29806	57129	80840
08	56902	15383	94277
09	69807	00954	88665
10	74385	02802	34044
11	77137	49479	86406
12	66354	97073	44152
13	01142	60345	77176
14	57152	43901	63159
15	48664	82295	16350
16	95563	94260	23066
17	14475	92332	52665
18	31195	51742	11212
19	83154	39474	98101
20	96853	86273	85331
21	50636	85720	82665
22	70172	37068	82250
23	04414	15835	35914
24	78389	09613	81894
25	94678	83389	15007
26	25159	80425	29013
27	37047	45296	72343
28	34913	32363	68308
29	30892	13329	89365
30	76282	76119	22166
31	87625	41239	23013
32	97787	52493	48217
33	31665	83755	32335
34	53557	41957	43210
35	12997	57989	13516
36	37299	40923	97001

	A	B	C
01	54654	94606	16951
02	49702	25827	25147
03	69462	17826	99712
04	89567	21519	23625
05	55216	16882	93646
06	61784	94435	04699
07	15792	25430	43933
08	82017	26392	76970
09	69093	68481	06544
10	86413	36132	26070
11	08672	22409	15180
12	73168	37946	00324
13	62282	23302	17932
14	00473	99794	51083
15	77410	91383	84180
16	41582	08727	63430
17	18567	64820	57425
18	28120	78997	49587
19	89890	45259	39981
20	24163	80284	20697
21	10777	79129	13534
22	69798	11176	71003
23	63291	10992	01727
24	69561	53181	37789
25	34766	22757	35527
26	20482	15125	70725
27	33044	54608	02100
28	73901	68137	83161
29	77150	59519	27158
30	79459	34827	08071
31	23019	46543	05384
32	37054	98950	79704
33	12922	37060	09253
34	31652	54227	74653
35	55056	06221	63768
36	78124	33701	12246

	A	B	C
01	74432	65069	00410
02	50145	34202	70565
03	29686	95066	37326
04	33071	05823	53153
05	86230	25401	72154
06	26076	92797	31294
07	12408	94313	98473
08	44556	91297	94170
09	00997	80652	49336
10	23373	19479	28378
11	71384	17843	44768
12	99104	15293	00285
13	36718	66628	57886
14	43637	79227	82843
15	75327	22027	21888
16	89585	13181	46573
17	33319	59115	92732
18	01608	81002	67109
19	68000	31461	45467
20	95258	97390	69007
21	57210	26516	36645
22	52807	13997	62679
23	83028	88745	98012
24	49545	39135	34978
25	94533	56139	97416
26	48064	98512	06942
27	49840	40681	22937
28	46999	86544	81643
29	56425	46756	33829
30	59573	25910	09737
31	60399	10188	68065
32	61319	89107	64608
33	22009	02744	27497
34	51509	56669	24726
35	96502	69394	62562
36	97466	27541	20733

	A	B	C
01	44328	93439	52918
02	65224	59917	69571
03	24257	14326	24676
04	46823	10691	75113
05	58800	50844	98831
06	60645	98960	34162
07	19494	14435	62926
08	55101	02815	11146
09	86771	29144	40267
10	61252	81471	06450
11	36511	19972	33316
12	74918	24456	13015
13	71663	61627	94598
14	45340	02193	30131
15	82254	87611	51291
16	05634	73165	36026
17	85520	99336	12401
18	45635	03753	54172
19	88489	33898	76572
20	79521	11307	94542
21	64698	16345	54570
22	48458	23688	33259
23	78329	86235	60385
24	96387	62507	49890
25	81398	09449	07172
26	89563	89190	59064
27	31202	43002	66734
28	38146	14154	41124
29	21618	56818	40902
30	60615	86381	20660
31	20904	24984	04296
32	45306	70614	34598
33	69559	91772	40169
34	87445	68005	05264
35	35618	07348	13003
36	36226	20397	90271

	A	B	C
01	14607	91916	52226
02	81870	33064	75782
03	78762	90889	37901
04	68935	92758	63564
05	41542	98970	82250
06	78962	50530	93889
07	05335	52641	02599
08	64067	23766	00826
09	83249	97011	06641
10	68861	65333	56045
11	03994	30550	98206
12	02843	73024	98683
13	46862	95315	24239
14	02368	20984	68384
15	89106	48723	25405
16	55139	34665	88442
17	59605	02371	75772
18	21603	83938	79858
19	06650	73631	54531
20	38295	38853	23969
21	77492	59975	08926
22	91626	41192	10956
23	08472	25363	06380
24	44216	60874	17277
25	24454	05370	16127
26	83979	96190	10037
27	98825	88616	10103
28	18489	16821	16922
29	21460	79266	79438
30	71136	78957	15150
31	72611	81275	87343
32	26773	67110	25654
33	44954	68984	70588
34	29671	90327	21450
35	83436	86701	40942
36	79458	60278	78828

	A	B	C
01	01930	81770	75837
02	02313	55825	16399
03	25916	17785	88489
04	53416	99108	22059
05	45708	05993	67833
06	67102	77456	66836
07	52795	53252	03063
08	07794	73397	55329
09	81689	52502	97466
10	72786	99244	97841
11	42005	54812	29605
12	37554	14197	33087
13	77218	32620	74034
14	29434	11631	76327
15	81148	06630	53272
16	84232	06751	72950
17	87035	81065	81563
18	41761	19208	43975
19	73669	85881	04139
20	27886	28180	04756
21	03532	73004	00100
22	27624	80818	31783
23	41460	05017	37855
24	95950	62068	89535
25	20241	91560	34583
26	18506	76823	90949
27	09909	63367	65662
28	08586	17754	70612
29	53510	38699	05944
30	22075	12810	69050
31	61473	89604	27663
32	73057	89485	26097
33	18892	38120	80247
34	59673	68812	72649
35	62955	57643	85824
36	18499	89812	29519

	A	B	C
01	02102	56331	15917
02	66734	41387	24541
03	97115	39584	92691
04	31897	57904	51421
05	83758	64584	72595
06	96102	41399	47819
07	38765	41159	64219
08	67226	09350	92931
09	89554	00433	86202
10	28725	83838	02003
11	18348	82854	39004
12	56884	73317	38080
13	62761	17331	29504
14	22498	82874	86345
15	53258	08596	94042
16	63114	81177	10888
17	17420	55424	54280
18	87663	50261	30646
19	34310	39379	41071
20	73260	71051	31379
21	68090	02066	97471
22	41961	09187	49171
23	54137	84009	74314
24	04349	58387	37450
25	85138	35578	30173
26	45268	45406	34720
27	35707	47818	94085
28	10605	83611	98653
29	39724	90543	51303
30	17758	21620	30704
31	12705	18349	26165
32	77421	72759	20340
33	70991	25618	42126
34	29570	35404	06367
35	10178	08124	35546
36	55858	46798	88362

	A	B	C
01	61835	52527	34849
02	55217	09837	83762
03	02100	62130	98891
04	72345	33816	52063
05	14458	95365	59747
06	59681	72982	09288
07	59763	87284	95689
08	82484	57154	46695
09	47715	67698	69288
10	37971	92208	20321
11	44863	28692	75609
12	18041	91632	65956
13	32193	12437	82401
14	60635	34400	01354
15	93626	95067	51756
16	01395	58252	72130
17	99159	58492	58776
18	94430	11474	64660
19	34370	28759	85592
20	19877	52782	95906
21	05144	23708	77941
22	76578	90699	86834
23	42253	93982	48747
24	98034	32576	24536
25	16680	97697	74375
26	21427	46267	36161
27	40154	73331	66766
28	58789	68389	39316
29	27460	89526	46681
30	18836	63621	33065
31	09699	43806	41431
32	55631	23251	04784
33	98906	81238	50510
34	55486	30347	63343
35	13665	13055	01688
36	35979	75621	96781

	A	B	C
01	26883	72900	90183
02	73370	45198	59207
03	98647	26202	92411
04	81130	03960	78072
05	82236	60818	23244
06	55786	31464	41886
07	85856	86339	33266
08	16761	25215	11042
09	29034	70229	88282
10	97313	16952	29465
11	05424	50631	36945
12	93644	69403	35405
13	79167	26016	15272
14	23135	28618	60783
15	97414	55786	26544
16	27635	65059	36717
17	02384	56732	39777
18	76964	98340	84507
19	23365	48904	12643
20	11251	79889	67459
21	72302	58702	49061
22	16333	40430	38953
23	24088	46329	18371
24	65564	36861	76836
25	09509	96590	15877
26	49261	82352	07662
27	41915	06368	61142
28	96766	84394	73067
29	39209	62377	48089
30	63144	61770	57329
31	23193	65365	79911
32	38757	36875	06507
33	36719	58754	58745
34	98102	86382	01880
35	68483	98764	64404
36	56485	34273	40808

	A	B	C
01	84003	54448	05907
02	19263	49237	20642
03	52105	23304	02678
04	73241	94167	08661
05	78705	16123	74728
06	55948	47930	81418
07	56334	60562	15654
08	01747	76832	23755
09	42529	32792	53578
10	19232	70296	72292
11	41474	50512	12987
12	56813	50949	74505
13	47457	10448	02943
14	87462	27864	15331
15	81532	62828	09009
16	01336	30997	88150
17	81461	87973	05989
18	72389	61747	88748
19	08146	50387	92530
20	96022	53312	69530
21	08477	51628	21340
22	82331	80765	28618
23	51868	09138	90424
24	65082	97754	37683
25	93174	77201	88073
26	47082	05720	40895
27	04116	30630	32628
28	24986	68381	12461
29	05957	70842	97858
30	39803	74421	20523
31	20030	97661	50094
32	44423	73243	78941
33	22086	59842	66392
34	33188	04077	20641
35	58444	93457	44613
36	37482	67494	58505

	A	B	C
01	66136	41390	62152
02	84417	21367	22983
03	30877	12416	78876
04	33540	99908	86979
05	44088	11703	84360
06	91883	39466	29410
07	36693	59712	89276
08	89828	16632	41907
09	43951	82028	85154
10	57897	99337	04049
11	22262	01969	96873
12	28138	13504	37703
13	91488	52611	49478
14	69960	58227	75429
15	29479	12439	61722
16	39438	59560	94341
17	27716	90456	72630
18	77946	76732	67104
19	28343	85951	40396
20	60957	18968	78811
21	87419	69369	57538
22	44870	15657	05459
23	90934	68877	47672
24	21264	47431	67421
25	03537	77735	85998
26	68566	27068	62420
27	90765	14560	21291
28	19263	41591	03981
29	62367	55801	57738
30	69842	50670	88385
31	13334	95292	94951
32	55750	30942	84926
33	64290	87262	78563
34	71225	54863	99918
35	34821	70747	68204
36	77768	48721	46213

	A	B	C
01	90471	65553	04661
02	36037	53743	55580
03	69790	93924	83861
04	91012	71535	80882
05	02681	29980	80244
06	25460	41044	78172
07	61483	21167	36155
08	43583	58860	42255
09	91007	02151	63483
10	55550	26320	15380
11	11674	16007	67324
12	69131	98407	91431
13	54670	80948	64613
14	69383	88345	54503
15	65615	69136	45153
16	76873	54327	93120
17	33198	64865	60547
18	59027	57091	56331
19	56951	57065	64360
20	52990	53796	54728
21	02773	29245	33824
22	66314	25592	22469
23	92621	82523	50505
24	19615	95300	88877
25	60715	99914	59598
26	43997	01423	20007
27	20348	34263	96626
28	17778	79679	54858
29	20524	29871	28949
30	93765	82388	39656
31	70140	09795	42930
32	87263	73199	01709
33	38169	01420	82244
34	58274	37589	19544
35	84165	64393	41923
36	00155	55201	59713

	A	B	C
01	67774	48221	99418
02	93601	90409	62628
03	10759	33510	99915
04	51561	57258	21943
05	04793	52040	64509
06	77463	41733	35931
07	51185	39715	21144
08	85720	43463	17745
09	15146	75108	77802
10	18822	50261	77688
11	52498	29073	56338
12	58267	22831	11755
13	50710	15832	87726
14	47707	07345	87757
15	91907	60697	81098
16	95221	13647	87591
17	24434	86607	16185
18	26333	36986	29992
19	12447	60538	80469
20	73891	78411	58189
21	98171	07216	35705
22	30415	01023	85615
23	14283	05425	48949
24	41782	71421	24845
25	70992	88098	72837
26	83522	89391	26337
27	09295	63340	59395
28	16877	34460	13875
29	85162	79274	42240
30	03685	73911	87202
31	00053	00307	43927
32	79615	48741	07216
33	44321	47705	50511
34	24534	30764	60278
35	86720	23092	43782
36	19537	31734	67446

	A	B	C
01	26895	12400	35701
02	62043	47278	63590
03	37134	87779	97513
04	73790	99198	71438
05	66551	05321	37012
06	48235	09794	17232
07	34463	65684	16677
08	29806	34097	22533
09	45951	39745	94276
10	10143	42751	81676
11	50725	18213	05063
12	85960	90929	79168
13	56891	44584	85266
14	12375	81263	77243
15	78402	03077	31813
16	90405	61870	71189
17	31019	00391	55071
18	52437	11126	36918
19	21634	82617	82462
20	52487	12847	81278
21	98988	26835	91358
22	91605	56023	28360
23	30789	20888	26524
24	78098	34234	50666
25	63526	66807	27833
26	31062	75635	66929
27	40749	46596	31227
28	30669	98014	04603
29	12665	39443	82024
30	69904	47291	08712
31	52223	81471	91111
32	41715	55662	87141
33	64982	84420	17194
34	66948	27145	56779
35	60966	30301	58132
36	60201	61468	85955

	A	B	C
01	82709	84079	95068
02	32759	94000	09103
03	03703	34107	04137
04	09195	65024	81427
05	45832	60935	07882
06	24914	79411	51989
07	36472	54446	07946
08	98429	93699	25800
09	76203	12903	41088
10	15434	58518	64529
11	46663	69758	35927
12	64670	60054	54478
13	86037	47538	22524
14	57697	96964	81470
15	78895	32730	20640
16	72210	86919	08607
17	00609	64519	55765
18	38964	90537	96233
19	25275	42323	11398
20	60600	63057	79000
21	29220	41587	25395
22	47483	35483	66486
23	24217	32412	21976
24	45770	18468	37070
25	03516	19048	79222
26	13804	75961	03625
27	88205	85234	06687
28	23101	02534	52960
29	32762	89732	77722
30	14250	27377	47952
31	26272	68859	66640
32	68603	14494	59885
33	11946	04871	06750
34	86791	90504	01536
35	80254	67350	34202
36	98407	06365	61164

	A	B	C
01	79101	88478	85385
02	58548	32140	18773
03	13179	29765	24888
04	63511	59102	68477
05	27896	97377	29017
06	95888	61970	61825
07	27176	68063	92351
08	87482	78752	99475
09	01651	58147	17990
10	24933	36316	22699
11	81623	73072	21362
12	48174	27758	57116
13	71827	70734	59682
14	44033	31726	24040
15	03736	18746	53753
16	00002	73346	54837
17	31079	22363	03802
18	08722	70367	68009
19	37530	08442	46389
20	96452	88056	29213
21	99425	96725	29765
22	66559	66478	31326
23	31210	28105	99358
24	65837	52046	71733
25	29152	79854	94297
26	01871	63702	57446
27	25786	79032	91439
28	48284	79694	05929
29	53315	36028	00666
30	44599	78078	14596
31	95461	83090	94218
32	67289	75471	65465
33	10683	97859	50415
34	10980	21442	37081
35	81832	91230	79795
36	25200	07098	68779

	A	B	C
01	24903	34012	92300
02	50358	68916	91566
03	48131	35925	21535
04	11742	99250	92554
05	77306	68018	31904
06	73100	53531	22590
07	14407	29818	33524
08	46867	05807	83758
09	88720	52267	04180
10	43403	43324	17669
11	95264	85977	60814
12	60083	73264	66588
13	04983	91687	60565
14	84685	34947	62096
15	66696	08348	82737
16	32433	67774	75441
17	57235	16395	41460
18	89464	28240	83046
19	82307	03049	25499
20	11694	13620	99217
21	20102	64768	76372
22	87437	30124	35942
23	05475	19046	11907
24	59350	32437	75775
25	50667	88791	00243
26	86022	56781	73461
27	34911	84055	37019
28	62092	02164	84560
29	10815	04743	92925
30	36220	14659	63915
31	72705	69920	75312
32	48816	51559	36437
33	19669	12551	36893
34	20403	75061	59911
35	74975	10034	92934
36	67504	61033	00859

	A	B	C
01	21958	67739	28420
02	12884	39269	15869
03	66253	17737	99185
04	25291	16543	46680
05	25755	43254	47839
06	01696	83541	50093
07	29230	39504	92200
08	55178	23204	36177
09	93733	46356	29085
10	62668	76900	13277
11	41639	41149	26133
12	33801	89504	19902
13	93717	63213	69098
14	19820	75119	07590
15	87454	01999	94740
16	68116	68080	54196
17	47062	01851	53509
18	64429	39910	97961
19	47896	16821	34056
20	71200	32944	35924
21	43052	63077	92452
22	96076	96879	77981
23	11972	55575	92342
24	05539	79649	19082
25	35344	84052	00837
26	38017	95412	73653
27	74506	51687	27020
28	39241	29998	98974
29	41604	94895	95173
30	79826	02679	59138
31	58644	41482	30650
32	13703	77552	37189
33	20563	94003	99367
34	82877	07696	65159
35	87804	08572	93423
36	94682	88690	08544

	A	B	C
01	46536	74916	89135
02	17877	07964	48107
03	49505	35270	88125
04	47109	51922	68872
05	39621	08238	28433
06	38445	88873	14437
07	86587	17341	49991
08	72015	09711	81208
09	79967	06342	27594
10	71944	71098	51658
11	74886	86819	49933
12	93756	72581	65628
13	87399	25820	90496
14	61767	94348	88038
15	66765	45115	37719
16	60786	59161	23784
17	47852	87554	68737
18	80820	56195	62295
19	49587	42108	55689
20	66225	39674	32244
21	23351	77418	37658
22	14868	82892	33586
23	91166	64504	98833
24	21368	73342	48488
25	14787	19237	14380
26	33279	03109	14214
27	83201	50968	01219
28	31256	36102	15114
29	59187	37324	52698
30	97195	82117	71029
31	67959	11313	52420
32	06933	73161	63455
33	81439	90396	19126
34	40106	92136	11857
35	76717	29801	14289
36	87220	81827	50396

	A	B	C
01	72134	48860	28257
02	23475	29383	69091
03	42388	45669	36950
04	98818	92486	08748
05	50671	96045	24188
06	34456	82455	64096
07	23538	89143	44653
08	51925	29228	86707
09	03640	29716	64853
10	17711	06434	26812
11	97631	94841	96744
12	58247	30670	58814
13	17539	29621	72232
14	28697	57955	28847
15	09616	84578	00459
16	29729	14306	08066
17	91097	28668	88205
18	52210	79410	42081
19	28235	33817	44812
20	26538	38348	33355
21	62270	75561	37373
22	36553	34409	12135
23	09034	39238	10217
24	24608	01154	80218
25	71251	55919	53945
26	62927	11295	33852
27	77724	37236	89560
28	85137	80174	58870
29	82670	00681	53839
30	91357	65611	17572
31	69090	86361	94257
32	27913	78378	69362
33	38345	31785	66897
34	54782	84735	90319
35	99466	90643	06705
36	85273	20469	86174

	A	B	C
01	20779	40854	94515
02	54543	26678	63860
03	61187	54243	56908
04	83415	37485	30972
05	41293	09456	10340
06	22561	02336	60882
07	68166	97317	09740
08	11174	98666	16388
09	90035	01979	49613
10	27762	14394	51067
11	66632	25325	18852
12	06378	37082	82568
13	43515	03232	72908
14	55943	76431	26352
15	90842	53061	24241
16	45034	71820	14349
17	55486	16711	43013
18	40332	85988	43195
19	61709	26849	34495
20	13046	52954	03431
21	70813	16663	64809
22	72121	23121	64929
23	69891	77532	40491
24	00179	52204	44666
25	42594	92785	35351
26	21212	71020	16593
27	63284	97612	93876
28	55600	98797	71370
29	69955	95127	83657
30	67658	46937	87864
31	12406	57467	20505
32	96984	18455	84977
33	30060	70594	16663
34	67817	37390	35970
35	24626	76069	68896
36	40991	89484	83903

	A	B	C
01	57392	02210	68086
02	12193	00127	31037
03	29752	52158	17662
04	86526	76320	47953
05	78132	16416	39068
06	98848	90006	32590
07	82536	00331	31547
08	07415	93260	52080
09	17803	82269	71618
10	85948	31972	10943
11	27169	20858	89045
12	32083	77981	71265
13	83735	99569	17960
14	02428	79362	04103
15	86569	82027	39768
16	30441	78579	97102
17	86916	76216	14880
18	11386	96979	51582
19	32687	24063	63021
20	03639	29435	63744
21	42066	60801	22980
22	74721	58597	19886
23	04119	58130	86122
24	13076	58879	71589
25	83413	71127	00976
26	33538	09255	83024
27	76562	70650	14958
28	48020	32483	55520
29	40032	37751	71519
30	57252	94346	79224
31	93128	46236	99212
32	42562	96913	95381
33	75489	07561	62372
34	08383	60673	97772
35	04756	46151	33290
36	10692	97336	17563

	A	B	C
01	26738	88985	74810
02	34498	92168	93800
03	34864	17765	93399
04	98770	32925	18323
05	59805	02196	10787
06	94112	25713	12733
07	47340	46718	11814
08	44736	81118	70767
09	18222	95635	83177
10	42166	54967	60233
11	77150	62227	64022
12	55300	46579	88629
13	06575	59805	00362
14	00151	08556	28897
15	18348	85184	99762
16	94221	54852	33946
17	43756	53271	23217
18	92195	11298	44490
19	97815	58578	69256
20	88168	93451	66510
21	44250	97322	53325
22	50262	70727	26326
23	26996	98428	21592
24	78891	38801	48594
25	09318	80922	05576
26	52938	76955	59506
27	98252	32008	99735
28	29526	87227	36076
29	93150	12775	36521
30	48235	72087	35739
31	63832	82278	36574
32	37932	53420	32214
33	94609	03045	78134
34	07594	79789	28915
35	76611	26436	01433
36	73806	47538	09687

	A	B	C
01	03097	24818	52828
02	68376	34190	18345
03	67664	04910	83401
04	19198	09288	66427
05	33585	81234	83469
06	52339	92756	82110
07	19186	04190	26579
08	66854	74811	66607
09	70199	85800	82043
10	03123	75059	14635
11	06944	59892	43379
12	89980	67369	00288
13	13145	88706	96853
14	28229	73031	36660
15	20789	14315	25925
16	13073	84257	05877
17	33839	10789	42340
18	92102	04662	95134
19	93684	77584	40458
20	57568	63221	82598
21	24544	27782	10594
22	74105	40792	71156
23	84499	47515	70823
24	85251	01731	98839
25	56927	39065	38972
26	42206	01130	76713
27	44177	50030	38074
28	13938	39579	68878
29	50749	37734	24989
30	88796	41120	45209
31	04351	97822	06740
32	82830	21803	32171
33	32534	65617	68082
34	86354	05038	96901
35	04417	62688	53298
36	05294	76149	70873

	A	B	C
01	30036	18123	94235
02	73318	17477	13568
03	60439	98566	81771
04	34709	89510	04401
05	89169	42985	14848
06	97129	06293	60700
07	34888	26320	56618
08	80517	32808	61836
09	98002	21567	49483
10	14865	58061	63440
11	95382	75380	89535
12	88541	18050	35765
13	78024	72142	38265
14	10174	28786	40223
15	79627	45886	56086
16	64742	17782	69169
17	19172	69085	85189
18	52147	06074	65669
19	08769	76532	76563
20	42989	29184	44192
21	24332	19963	62074
22	35506	24698	82016
23	79495	60007	42627
24	41919	39752	95956
25	11993	27587	98540
26	22622	92948	92873
27	27054	19882	91555
28	26694	18587	96519
29	82371	10036	23851
30	58239	99955	94681
31	87242	94964	58682
32	57851	86639	12812
33	65931	89394	07529
34	70105	15593	88671
35	47130	72759	38171
36	49202	02783	52042

	A	B	C
01	71435	38829	73027
02	78939	62487	83956
03	59428	50333	81808
04	49857	30625	22251
05	24892	56764	68724
06	55310	08719	30672
07	62864	20904	00686
08	33224	75068	79229
09	22454	00639	09965
10	89144	26809	49038
11	76172	69527	89142
12	17745	67442	87581
13	10058	84114	22700
14	19267	48326	74583
15	38054	72515	71583
16	33799	17002	01036
17	16284	11876	96964
18	58079	94556	62238
19	47950	78486	78307
20	36884	39551	02241
21	50755	52708	35155
22	15739	06062	80720
23	90560	19677	76165
24	53739	07264	48135
25	88878	44213	59216
26	85716	82276	78779
27	65990	69054	94481
28	69811	05668	42221
29	81103	81453	79802
30	97074	81942	22086
31	68675	08646	70772
32	37335	36772	15380
33	13838	73498	35187
34	58399	36888	22578
35	89544	48315	75466
36	02589	46590	17802

	A	B	C
01	48324	58065	14159
02	83160	81141	87093
03	24397	64161	67428
04	04686	20794	45247
05	21492	41638	22376
06	81763	41614	42175
07	97587	78896	72978
08	94717	74975	96558
09	66299	02850	36601
10	68357	79480	38336
11	92814	01150	02854
12	39100	96133	65477
13	37400	93200	67271
14	46820	70073	73652
15	13566	22320	00620
16	20461	56956	16850
17	18736	29476	88425
18	37101	92070	68425
19	29000	35204	49975
20	04460	83258	81430
21	76961	79712	91035
22	40221	84668	37816
23	24813	52555	82284
24	55898	56163	39342
25	06229	08289	80688
26	64033	79902	01393
27	60020	12797	47907
28	57889	72016	37699
29	07890	89350	63064
30	67741	73775	32277
31	47485	10881	49192
32	03830	14957	45344
33	76617	44928	93304
34	80757	64364	36236
35	22781	60171	14097
36	72947	98205	75391

	A	B	C
01	70618	95944	18881
02	58468	15062	38473
03	80727	47639	48656
04	21497	74658	01145
05	42640	18818	19910
06	57121	71597	57588
07	58091	49842	95360
08	93409	00913	56258
09	08037	80952	57148
10	73486	84731	05206
11	90864	52376	40222
12	02194	86550	49469
13	16543	52129	05970
14	11618	37453	63319
15	44966	21121	35491
16	25644	88652	07752
17	16491	78261	97823
18	08688	49422	16477
19	25734	66932	93982
20	83599	40920	58265
21	77530	84874	07771
22	02081	71914	07158
23	66059	87336	34483
24	14427	24170	55893
25	64758	83103	99828
26	52121	00202	90416
27	73168	08769	84272
28	74858	52483	88094
29	87001	59248	65370
30	58773	39064	73941
31	89952	70060	14922
32	81825	45837	74423
33	21353	36806	53295
34	66994	41461	40654
35	20709	27166	62519
36	87307	71663	25204

	A	B	C
01	83664	60822	74572
02	37418	97382	44524
03	96609	73618	33333
04	63337	22359	44022
05	00283	78647	07135
06	06766	96990	01463
07	20548	39335	73469
08	86925	44425	14455
09	26116	59370	70340
10	46597	10835	98291
11	21637	23440	43171
12	38577	01122	46145
13	70242	28024	04625
14	15098	77328	26484
15	25028	18329	11475
16	94739	92206	65097
17	10823	82607	69248
18	90673	45488	40618
19	18709	35448	37838
20	28786	98966	58605
21	67638	86781	25049
22	37096	57913	68652
23	06040	98176	90735
24	82675	76268	60662
25	35282	70806	63616
26	76494	77411	12433
27	50832	85126	27797
28	90538	31830	23770
29	87202	03856	92549
30	02969	52571	33171
31	21505	45236	17993
32	84518	13348	40597
33	77018	62826	06991
34	69813	58585	64289
35	24651	79552	63455
36	18424	61907	00117

	A	B	C
01	79940	74323	76918
02	62060	90240	95056
03	29719	69103	21402
04	33143	37685	15715
05	90133	84004	11659
06	66251	46371	81393
07	83750	22142	55191
08	92690	44281	28152
09	86313	31802	61247
10	23788	33412	01853
11	54835	19257	84286
12	10677	44284	95433
13	03212	41970	48810
14	29877	33196	28606
15	05141	73746	74744
16	98363	68535	15975
17	98006	99957	39076
18	66200	15803	54924
19	25756	99295	84234
20	02049	53356	94598
21	49721	78109	21746
22	94711	93942	62966
23	56987	48428	57255
24	65337	82901	06676
25	49068	31470	94874
26	79337	50555	34222
27	85360	75543	14763
28	76210	83031	78089
29	26568	47268	03944
30	14602	86984	78316
31	56471	97276	40386
32	21835	43970	24786
33	04145	14491	44548
34	99233	78346	33252
35	95046	04691	38838
36	31068	20913	13034

	A	B	C
01	93296	84723	36132
02	96835	37453	28196
03	08352	07565	50928
04	50939	14024	00970
05	80656	91486	50182
06	25529	06094	38224
07	16693	38533	07673
08	29360	95571	60196
09	31753	97438	80377
10	50811	21605	31684
11	83662	77279	15164
12	10893	76273	39060
13	55294	45531	85464
14	32851	70429	49106
15	56318	17259	03981
16	48154	47129	32706
17	68170	67334	19895
18	92707	43844	35395
19	66221	56664	82799
20	65244	95194	04902
21	73176	83583	49524
22	33908	11994	79042
23	00767	52072	16099
24	91082	39055	25620
25	11725	95027	54400
26	06052	78103	59745
27	18528	04431	79377
28	11559	54329	57529
29	52250	41672	13612
30	51824	61413	57091
31	27037	85438	69577
32	44340	16981	91496
33	40915	77588	99382
34	66367	66080	77896
35	01694	12156	23510
36	09947	23289	58798

	A	B	C
01	12036	75160	32908
02	41869	89892	93646
03	41235	82208	91955
04	92158	98353	26669
05	63215	44812	29722
06	57421	68333	35833
07	62036	83757	73139
08	74552	85730	33490
09	18341	66709	95969
10	56112	76551	95787
11	05261	56945	01430
12	73639	06273	03467
13	80536	77854	55744
14	96490	48711	11137
15	69620	49968	44286
16	69297	73466	31668
17	84226	77992	99924
18	08507	41201	67599
19	05908	76690	44155
20	14838	91743	60653
21	47785	62223	38981
22	17435	73615	23208
23	07747	76627	69763
24	40961	07314	45818
25	48681	18859	71404
26	52036	30431	32789
27	41822	02950	02697
28	91058	00415	48487
29	52597	57814	24952
30	22866	05318	73642
31	97123	74286	23466
32	32275	67184	44918
33	78401	79039	86664
34	21937	73933	33324
35	26419	14734	78917
36	97606	96924	08607

	A	B	C
01	56723	69132	13299
02	19415	87133	87858
03	09050	72380	61792
04	70539	14894	39363
05	75775	98239	85371
06	39507	65901	54050
07	96597	35707	18695
08	97664	83229	60777
09	52860	63032	37379
10	03835	19234	97784
11	33167	07079	81844
12	03098	98964	67610
13	87142	52406	35616
14	51391	05411	36017
15	97511	18344	97411
16	25898	68151	93130
17	39463	27733	46305
18	47638	94886	57920
19	36236	26163	21130
20	95987	90254	79619
21	85548	14132	85127
22	54546	00591	07031
23	66855	32438	33261
24	47575	19817	16856
25	20275	88134	85821
26	14454	33299	22618
27	74706	64187	72184
28	17692	09269	88343
29	88639	83555	42669
30	02672	14168	62730
31	06467	47576	00414
32	84924	11574	78100
33	83388	93762	36874
34	00927	81416	28441
35	26621	45063	79167
36	75759	39916	75204

	A	B	C
01	56085	43892	42620
02	95927	45288	51044
03	85644	35338	45681
04	88453	74636	56363
05	55237	76767	62401
06	32780	75943	51098
07	69824	53260	36154
08	89594	94515	78286
09	89293	82708	63064
10	71020	32647	56538
11	44294	41614	02049
12	94416	33634	45535
13	10609	98344	15668
14	52185	87180	19879
15	06128	12971	85073
16	59726	74135	06304
17	59893	73050	21657
18	71469	09764	37810
19	49126	49308	71725
20	78038	58573	96152
21	67085	25627	35668
22	76839	17889	93629
23	78105	69514	63840
24	36728	29909	64586
25	01751	27500	55740
26	25355	39998	46408
27	45114	38189	02217
28	34868	97687	39066
29	89557	69173	38843
30	39285	54439	87535
31	96360	44712	92350
32	18164	28804	29388
33	17478	46042	69890
34	17245	80050	81165
35	24487	29115	76356
36	63754	67459	15540

	A	B	C
01	80339	51119	77970
02	06279	19207	89291
03	85814	98956	33391
04	32673	65589	53578
05	60041	13335	01590
06	84196	07719	91532
07	28827	38513	65759
08	62059	18029	02035
09	97553	84893	26348
10	92297	20457	49899
11	39070	14648	23042
12	40841	27417	72792
13	92169	82289	64744
14	70174	84591	69352
15	93035	81742	43883
16	13517	73763	31943
17	10545	57284	27596
18	32529	39495	11304
19	77530	72411	58470
20	01399	82541	50971
21	75043	69083	76485
22	29558	16690	22770
23	26624	36575	32998
24	63594	37891	08280
25	15547	56785	79050
26	82708	65029	27380
27	04569	34097	26449
28	54265	60451	53812
29	45149	99411	36775
30	65255	00725	53680
31	68339	44049	97454
32	50149	50536	23148
33	19219	62137	32007
34	63677	28780	37344
35	24864	18078	25788
36	72445	34966	16527

	A	B	C
01	14891	55492	16869
02	54795	94075	83457
03	92956	46534	94325
04	58364	34138	25217
05	14525	08265	43647
06	77750	95426	24194
07	80475	96825	64672
08	98484	45183	28074
09	62237	25103	91253
10	34752	70769	84880
11	43965	82517	50126
12	45760	85190	19785
13	91935	14675	69016
14	06517	74906	75681
15	52756	01412	13871
16	65590	26499	30305
17	73930	59736	57420
18	98268	03220	80607
19	15124	13052	79428
20	66211	05000	45734
21	18578	83884	85128
22	32103	84163	32073
23	50908	35686	48003
24	25727	67190	21183
25	86115	68697	07037
26	95263	43852	68942
27	08422	08335	32378
28	16642	27792	51994
29	73804	34624	84132
30	02030	71775	96313
31	29265	94605	75638
32	01219	05785	68637
33	43339	74766	40190
34	65422	29255	88175
35	40945	50714	16297
36	47015	61785	88513

	A	B	C
01	61014	75903	58817
02	24732	63004	42779
03	98316	57637	91355
04	14417	00431	15408
05	37470	93506	99920
06	03561	55818	54047
07	59654	66914	55150
08	93567	63731	03242
09	95558	24849	07891
10	53205	08410	26903
11	65889	48964	90771
12	60976	46157	81273
13	24329	77297	52976
14	56376	24318	17488
15	56254	79470	76426
16	68632	55840	93860
17	07718	77499	27234
18	31028	79962	11901
19	03398	41422	28919
20	18657	71023	98103
21	99117	94331	04817
22	46426	13352	23047
23	40975	28670	40027
24	59592	70505	53351
25	90518	70333	59019
26	69517	97036	24787
27	15764	31331	68944
28	94549	47658	89809
29	21734	54262	81014
30	47683	44839	30805
31	74656	82869	14841
32	76480	54249	46975
33	32851	48605	30518
34	67132	34589	60903
35	98129	26421	49129
36	06799	46054	71717

	A	B	C
01	64949	14751	99050
02	57910	64329	57595
03	84623	44532	92997
04	22507	42578	86230
05	30651	12903	84178
06	35075	32038	98351
07	51320	82847	17388
08	49486	02855	83907
09	54751	66144	79568
10	09248	86417	42586
11	09071	95828	36200
12	32206	31816	93908
13	79389	19682	29176
14	48605	86465	81028
15	74495	04578	71139
16	20078	72025	62966
17	19387	55922	13331
18	75744	65481	82243
19	61821	95939	48230
20	60535	37460	96303
21	70724	59478	51454
22	85064	97346	09303
23	19058	53368	83346
24	31609	50107	00899
25	29059	23287	99781
26	76181	83301	24777
27	86332	08336	07364
28	48121	90070	71738
29	06420	11409	97756
30	68264	17990	86535
31	62140	40147	89784
32	52011	80469	98768
33	29289	15223	75216
34	40425	25124	15120
35	55673	23240	19206
36	59545	70319	62072

	A	B	C
01	07094	00261	04239
02	98411	65652	59489
03	21861	34114	61685
04	25695	45049	99008
05	68610	32254	04448
06	68906	74316	97686
07	21702	85224	22963
08	00248	57564	76888
09	60083	90704	02380
10	48216	40779	84013
11	38923	28482	89313
12	73144	49299	89687
13	02717	59776	95248
14	58472	06298	72661
15	80099	83592	50063
16	69658	19928	66883
17	53900	84353	29859
18	84292	53897	41998
19	29694	09925	06616
20	13545	77598	92653
21	09456	12492	76093
22	66180	77701	82554
23	16854	84393	61076
24	72230	74774	54706
25	90446	48369	07280
26	77789	64371	42909
27	06185	86105	72721
28	57404	14272	27202
29	00905	96371	98661
30	97951	12254	73093
31	69604	84465	54098
32	79907	17986	68401
33	70472	50748	22410
34	22924	53146	61795
35	06901	89217	14774
36	39857	71485	45636

	A	B	C
01	31522	30759	91977
02	87262	76397	45259
03	59281	32875	35845
04	82754	37684	61787
05	71591	00271	38383
06	62854	29272	66782
07	31136	20521	82385
08	01045	45845	61450
09	83465	70818	91802
10	28606	00355	57695
11	62150	95738	84381
12	01392	97794	96189
13	87880	85684	79261
14	18525	75052	75269
15	25725	87430	98773
16	08730	70284	92556
17	10469	86177	59430
18	69657	60328	78691
19	15325	84539	94878
20	80513	21728	65377
21	79536	95421	60282
22	60039	88372	47026
23	30535	06251	74520
24	17769	27233	92989
25	35251	62867	27352
26	96524	69926	62963
27	79543	71087	23863
28	24071	31635	25291
29	35796	38451	92172
30	75112	43021	69119
31	19980	21945	79237
32	36679	69045	92209
33	49453	48036	80733
34	12030	73220	03510
35	01740	74983	90829
36	21775	01639	23263

	A	B	C
01	89197	82343	16362
02	62204	61682	04908
03	03742	56070	59115
04	56343	72063	43584
05	81056	98157	72777
06	07637	51546	99382
07	43565	67052	36344
08	53029	80889	33533
09	61636	51891	12177
10	04836	74368	22868
11	36910	47376	76293
12	35824	60968	80629
13	75293	09812	05563
14	90518	15117	02211
15	29914	28956	88579
16	70308	50691	27014
17	43576	71370	98502
18	54082	52238	25244
19	98782	05216	73024
20	72448	96698	81085
21	22755	64928	64996
22	78614	48532	91499
23	05220	49052	18510
24	82066	72274	61513
25	42758	71954	98549
26	74730	17534	60515
27	10998	66866	37736
28	00882	91422	11661
29	91040	99101	64286
30	62246	81693	15931
31	74760	39037	43023
32	04761	60579	80304
33	01216	73916	46604
34	62899	63174	26267
35	50956	01316	24779
36	50898	69590	29938

	A	B	C
01	80688	29950	16306
02	51099	87432	84583
03	64317	41355	51215
04	48764	52611	00355
05	36539	44137	63101
06	44036	72969	70903
07	92875	50524	97296
08	84189	95173	94828
09	22107	68688	24912
10	22784	05591	75927
11	15937	48990	19588
12	75825	34037	07281
13	11599	43268	14282
14	69613	14993	17619
15	18379	51503	06684
16	75072	31372	75876
17	67768	24717	95694
18	52174	82050	95688
19	84214	65479	80969
20	91162	79422	88276
21	77569	69555	02579
22	00844	75492	81914
23	32495	97258	37979
24	83516	65520	70577
25	31321	95588	96170
26	82326	89315	07292
27	36499	70261	33175
28	25413	58099	84784
29	51623	26456	52779
30	84036	92004	35785
31	50029	18288	41488
32	30903	78197	21937
33	82761	31111	83146
34	85888	85009	39416
35	76392	69780	01885
36	66629	00256	43699

	A	B	C
01	63996	92422	99954
02	47009	98969	38867
03	22027	27257	32559
04	92360	88948	65987
05	67255	86270	46555
06	26252	78984	27227
07	76865	70110	24868
08	89708	88058	48718
09	58424	29846	74121
10	99274	80912	29487
11	05128	99663	21374
12	00433	36826	32052
13	62328	78620	68266
14	60785	00635	88221
15	24727	65611	10248
16	80793	22396	27250
17	54530	60605	64621
18	16747	67427	86823
19	33062	39755	49097
20	42591	45318	88378
21	80567	37102	70411
22	43533	32117	87415
23	82356	09494	01284
24	74378	37478	04353
25	66134	30930	25462
26	62660	02824	50737
27	70084	44123	40997
28	03759	00406	35355
29	00627	23946	66029
30	65552	83584	52153
31	27097	36582	00476
32	98115	85851	95115
33	27773	82230	57621
34	14448	96853	98978
35	33516	15368	36451
36	66646	20844	17116

	A	B	C
01	86998	74147	55228
02	44839	77018	88676
03	92657	94375	65326
04	34113	84140	68406
05	28694	96822	29517
06	75878	42441	77265
07	99751	41519	34637
08	58247	05752	23673
09	65993	93455	37177
10	01968	89362	44627
11	21813	70012	85027
12	49339	52335	06620
13	70873	33452	44554
14	41112	66591	82467
15	63937	18933	46279
16	66658	04030	96505
17	60774	46605	11954
18	05593	08652	66665
19	52629	53556	71650
20	93551	49359	56017
21	13482	67552	13745
22	29699	91016	74856
23	89527	57902	94347
24	22834	64925	02940
25	54231	04185	24153
26	67927	01777	92994
27	29351	68801	20657
28	75962	78729	21658
29	33374	05793	96331
30	12992	30625	46074
31	10315	22216	59595
32	71892	26123	28127
33	98479	31638	28667
34	59864	73104	79160
35	55157	66271	31735
36	47433	12208	96978

	A	B	C
01	81843	67116	03432
02	85983	09713	11463
03	91895	14471	25161
04	39111	58110	62824
05	98768	22696	23427
06	40064	49177	70361
07	83457	81908	20614
08	99102	16281	50458
09	36308	13655	08411
10	10401	47825	36589
11	03485	59524	71391
12	79484	19645	62298
13	49291	79323	55150
14	24780	99282	39753
15	55031	65752	88322
16	13120	42385	59164
17	63282	41962	77936
18	89763	54804	96387
19	48903	88571	79335
20	04174	18727	97320
21	30325	82156	60530
22	49871	94331	76619
23	64841	95277	63158
24	65317	52502	25127
25	11513	79544	71225
26	08743	75687	08979
27	09950	07646	28280
28	55648	70339	56214
29	59819	64773	76325
30	82658	63053	46902
31	09243	82673	87949
32	96189	03171	06318
33	56888	90369	18614
34	79198	16270	99735
35	13154	52329	47250
36	22720	00806	64147

	A	B	C
01	39035	48955	59297
02	60229	05245	89399
03	73576	63026	73451
04	16074	06491	75031
05	88775	95311	51482
06	88270	45150	91001
07	87509	53791	53752
08	06411	43676	93015
09	90142	91410	31783
10	36992	17982	40670
11	27792	36612	85529
12	49627	08467	12641
13	75704	73026	18871
14	09096	86116	53216
15	97743	49990	39865
16	17562	48093	05255
17	41650	96109	81042
18	86436	76435	48127
19	47363	46524	53094
20	66720	98013	44173
21	51475	10557	53640
22	89418	46885	14448
23	57126	09585	28374
24	00685	79125	06961
25	86027	64765	48356
26	50968	44102	02105
27	57878	60026	32009
28	46945	76003	13102
29	54672	06142	54888
30	65531	38273	91125
31	83661	81685	26934
32	71591	43034	46660
33	16409	83378	81357
34	31367	01567	25772
35	18090	86221	42769
36	29115	03607	81944

	A	B	C
01	53656	64482	59299
02	20021	65263	23379
03	92459	43503	24932
04	72911	59420	03505
05	58521	30040	19824
06	59810	23697	26482
07	73656	53334	74290
08	85046	48631	11683
09	16421	77020	00835
10	88469	85797	94058
11	97604	82426	54839
12	31639	35279	67952
13	61855	04948	05817
14	58374	03155	80742
15	48934	71616	06504
16	28779	17924	44438
17	76937	37445	54453
18	34742	96586	87198
19	20949	79455	04453
20	48013	73671	25728
21	13263	06482	54924
22	18427	90094	37719
23	87928	11847	18120
24	29667	46199	91676
25	74229	79359	41895
26	12230	88291	02534
27	58848	09219	34837
28	19441	87672	01560
29	49923	22947	50175
30	39637	34731	32451
31	16472	65920	84528
32	16331	62157	47329
33	67259	59419	77669
34	62696	14725	38034
35	37873	11746	30786
36	20575	34701	86441

	A	B	C
01	95632	83661	51436
02	66996	66834	60354
03	77585	76740	49511
04	91475	71554	93233
05	63290	68249	92481
06	19628	15352	96507
07	92242	60667	58066
08	45127	28043	01246
09	29310	36409	48259
10	02527	92852	88283
11	46476	80931	39899
12	09726	15107	25607
13	69016	67132	65504
14	91553	25686	61904
15	06340	49553	17830
16	43965	77009	65055
17	35934	58175	71858
18	43468	03271	64626
19	56596	86908	40991
20	73714	09489	45677
21	54439	18324	93328
22	26267	32583	52881
23	69260	93911	28382
24	35485	94759	56374
25	59958	52511	84615
26	96496	13145	00267
27	33202	54280	24669
28	51179	88907	77330
29	00094	17393	70697
30	30172	95580	97677
31	61102	78343	83028
32	01131	29679	23859
33	47900	67522	40931
34	56097	59820	13978
35	10980	64260	06644
36	93178	89727	51322

	A	B	C
01	13896	59099	58415
02	25585	13062	57595
03	86755	92719	72258
04	60513	00110	11858
05	11828	26058	28862
06	32240	08229	95184
07	56373	64935	25354
08	92098	85996	42279
09	90639	44962	22932
10	55750	02741	57976
11	66243	39782	14565
12	27683	13121	61020
13	65566	28649	94231
14	71880	79880	44828
15	84200	64198	95853
16	12375	73643	27028
17	36846	69874	69388
18	82509	73574	99715
19	17664	29662	80945
20	56571	61714	26364
21	45426	26885	07020
22	77131	33462	05948
23	78333	67526	24857
24	54802	25563	03247
25	99745	62048	18972
26	17605	86431	62330
27	10443	82565	09675
28	07234	98166	08524
29	92550	49363	76818
30	67207	67370	33611
31	21807	67710	79294
32	39633	99647	48732
33	36287	04262	73900
34	60629	92366	52328
35	18001	48175	61691
36	73741	11571	71511

	A	B	C
01	02882	56468	85100
02	53841	16592	28884
03	28621	51379	38764
04	83829	10602	92738
05	11793	03458	09286
06	18205	10332	08320
07	71627	69664	05369
08	83128	54108	71296
09	96219	22290	17783
10	90515	65687	56549
11	22200	56910	54819
12	17201	92881	35847
13	83411	44248	85511
14	50768	99156	42264
15	62620	84816	90154
16	44360	97691	15584
17	34498	92906	63291
18	02119	95822	37493
19	82386	23416	57383
20	73432	66023	77712
21	86113	20403	91353
22	66321	48337	60945
23	48635	71556	57450
24	44452	66275	13592
25	32169	55037	66930
26	91968	47984	77832
27	13099	41893	83229
28	00956	32971	74503
29	03053	03645	05047
30	46906	80556	12410
31	53987	18609	85256
32	77726	00171	41155
33	83638	70534	76675
34	33718	92863	39660
35	99924	94879	79793
36	22169	19021	08637

	A	B	C
01	74781	44909	16416
02	92798	62829	47439
03	79090	27254	52816
04	38448	55922	64473
05	03063	89191	59344
06	88139	17302	84760
07	53931	70425	36650
08	41283	01883	69987
09	44001	77898	59277
10	89186	50563	94704
11	35585	69908	85373
12	66990	97264	95638
13	89482	43200	76736
14	20385	19829	46017
15	51183	98012	87749
16	64365	17950	34454
17	89020	25531	46863
18	86861	74764	11995
19	31874	48527	99315
20	45013	20652	71132
21	25334	51613	13426
22	50398	13308	52810
23	75574	87091	15895
24	18228	31727	93869
25	85647	45138	38749
26	71157	54068	79458
27	19792	93331	77120
28	22735	70343	43601
29	88143	72764	64274
30	18143	66012	68176
31	39375	01456	54686
32	45139	74475	98862
33	07930	11117	06265
34	24434	36935	71382
35	28620	96967	91867
36	02114	27456	44849

	A	B	C
01	26305	28441	68472
02	87602	52521	61979
03	11756	43395	13904
04	64042	61310	34435
05	84395	32985	82992
06	35223	00455	66878
07	57748	58450	99863
08	62082	37336	94504
09	89132	33824	88876
10	51390	40210	79603
11	97195	04000	84133
12	20067	29743	98573
13	12921	76739	66176
14	62805	30829	43672
15	29511	25328	25399
16	45298	00827	00355
17	56559	63724	81031
18	55926	43208	82784
19	35236	48629	20000
20	93373	44937	71296
21	75797	06567	16327
22	92918	95417	96750
23	69809	93641	20789
24	10962	32475	29508
25	29279	25527	26001
26	61314	22860	07612
27	02632	12908	20295
28	20164	48114	54778
29	62266	11392	78686
30	20598	84465	42783
31	81253	30754	19075
32	40792	67511	07049
33	00961	24488	96206
34	80463	44031	57397
35	35280	07810	23747
36	49021	77656	84052

	A	B	C
01	22044	89025	11411
02	45878	67337	63168
03	97487	93114	38474
04	30990	90780	98919
05	77860	83018	70503
06	16959	15942	08964
07	48121	33532	58435
08	75131	06384	78434
09	18989	06860	72451
10	27416	07597	41881
11	60038	19943	32944
12	54041	81085	92555
13	76659	95743	36071
14	09493	20115	41017
15	01102	84303	87562
16	60084	62396	58789
17	52662	56856	00298
18	34068	09576	92841
19	04717	32865	18054
20	11912	01535	52315
21	96124	77055	86398
22	26068	20608	00612
23	01030	07541	12650
24	43211	11537	25925
25	59608	81161	47581
26	67086	56428	70845
27	55783	95142	00287
28	69323	02133	67979
29	78474	13862	36705
30	60947	08629	70122
31	46794	36996	35541
32	16840	77409	01104
33	01571	01824	81340
34	05777	31142	24843
35	38204	71462	37187
36	32359	02905	97920

	A	B	C
01	43654	35980	74283
02	30748	13284	95455
03	81552	64459	32406
04	38517	86902	96188
05	24736	36927	77692
06	34114	21235	24831
07	41902	35725	47495
08	41422	25034	05342
09	21654	24344	27367
10	80105	86065	05113
11	09161	17172	10458
12	63068	15043	99237
13	65796	84980	12527
14	09959	53944	92592
15	20780	27078	28771
16	17206	65051	60820
17	25892	57586	67310
18	43286	26329	04662
19	27248	45552	77071
20	16122	07301	02103
21	94401	33152	43375
22	46968	73583	50068
23	98811	75846	77829
24	69874	54212	71556
25	45189	16142	45006
26	61890	17882	50446
27	97586	35631	88639
28	02066	67816	63920
29	40178	29097	72848
30	03346	64981	25936
31	37421	03565	99091
32	94016	46743	41836
33	17021	84308	37028
34	42512	49436	66593
35	20311	90154	85850
36	60605	35710	31914

	A	B	C
01	54608	02123	89752
02	21528	30716	75971
03	04216	88628	44735
04	06693	50028	24263
05	90581	86801	92582
06	62323	89702	28464
07	59680	17760	62959
08	46454	32462	33948
09	69240	66605	94602
10	96200	25240	73016
11	50109	67140	42046
12	18812	48868	22103
13	14906	69570	02007
14	48855	07133	55449
15	65327	14228	65530
16	60614	37895	74461
17	51491	75013	76363
18	14901	31206	46536
19	88025	11485	67646
20	76755	62859	39207
21	31062	28370	82499
22	94796	73738	45951
23	40788	26363	75186
24	28488	88732	35833
25	86953	14874	21753
26	37103	54538	39433
27	46332	12766	13962
28	03517	09613	74011
29	39410	61862	71911
30	71517	42696	95812
31	09922	06151	01745
32	89280	49825	94036
33	76395	66326	36442
34	15443	08293	19322
35	37978	14231	02191
36	96183	80413	00023

	A	B	C
01	81826	54918	08481
02	48111	76557	69632
03	14964	87494	12353
04	34360	31751	03751
05	74394	58369	53478
06	57459	36955	40004
07	25843	54020	80835
08	54867	13765	88228
09	36250	52131	38738
10	32864	48625	48630
11	07002	95486	81112
12	06963	45221	40466
13	25789	04830	21958
14	97399	85309	86641
15	07234	14875	01192
16	02935	32975	98232
17	21649	60148	28457
18	34722	58042	60509
19	90358	03493	31199
20	14559	37256	17249
21	92617	74343	19680
22	45423	73289	50852
23	98198	53612	03526
24	77971	10458	89683
25	70253	37278	56144
26	13312	48617	68012
27	33855	89496	56719
28	89400	39950	30326
29	89338	18331	28195
30	69508	69845	52015
31	77852	65960	34910
32	14401	51505	82965
33	14638	42124	84960
34	07202	65526	13096
35	81184	32704	97838
36	33269	46134	09310

	A	B	C
01	19463	60246	51223
02	03791	74211	94281
03	24659	50670	51106
04	29052	54368	60026
05	63523	31373	45738
06	17167	18050	79303
07	20924	03950	88908
08	05015	78875	80557
09	39824	33232	89657
10	57314	42438	76482
11	00789	43609	85488
12	73590	43312	93179
13	98531	57383	42245
14	68458	51030	11868
15	04517	94444	75179
16	83804	26008	60841
17	02708	32554	51130
18	05206	32859	10262
19	08616	46354	88005
20	43145	01855	68407
21	99346	05581	84093
22	97473	07976	10056
23	97093	19185	64827
24	73950	46234	47778
25	32264	64356	77069
26	04818	95424	74655
27	36763	40337	69045
28	28529	78217	11249
29	88913	63883	00522
30	63363	55055	95505
31	57640	97628	07379
32	26623	25217	57651
33	77151	16551	11586
34	27579	87252	44814
35	96209	67570	10632
36	86306	51969	61933

	A	B	C
01	66218	29532	11437
02	33830	33924	37940
03	13309	43532	45250
04	71618	69522	28901
05	72430	59810	43744
06	69877	56123	20900
07	79913	40179	56315
08	65428	86612	95088
09	38476	48428	52087
10	38617	55671	12711
11	47751	75561	47224
12	46861	62993	73152
13	07703	26804	50913
14	00068	71361	64353
15	28175	71665	23503
16	54116	96118	34569
17	31299	10245	86912
18	12262	14840	33555
19	38072	63424	28431
20	37237	08997	15461
21	50939	26674	58917
22	46103	97273	16535
23	23747	66841	31183
24	42229	72091	13541
25	92765	90392	13862
26	41576	09247	39870
27	13405	13104	47821
28	23102	00800	55046
29	92312	46545	84182
30	68159	89350	82417
31	60599	92571	29140
32	88014	45037	75760
33	15018	21643	86507
34	81740	12849	48926
35	58075	08841	13509
36	50834	31883	95766

	A	B	C
01	38837	28970	36439
02	56186	95650	62730
03	43824	35028	80492
04	46727	38191	97780
05	17960	88524	38959
06	91079	09022	27649
07	80621	97755	55020
08	40159	99660	14703
09	97654	79652	82291
10	56381	56151	83289
11	98550	46103	20589
12	41816	97408	16931
13	16090	60059	77055
14	84624	36561	72607
15	41466	20405	45193
16	50889	32227	32844
17	85485	93842	19772
18	99507	41230	16996
19	88300	71140	43652
20	22828	15559	20390
21	03977	26850	80816
22	66345	34569	87107
23	73898	62168	49087
24	34436	92821	09800
25	67135	29751	26592
26	52839	22947	18782
27	37727	31332	77360
28	71914	82667	47882
29	20657	48594	29543
30	25150	95717	63747
31	80666	33557	16195
32	43373	22613	70438
33	95525	93494	17639
34	15006	36672	84193
35	83233	32803	21965
36	32391	95945	99841

	A	B	C
01	82580	36802	12779
02	98987	66513	27503
03	40387	49136	87469
04	57160	46173	89678
05	58280	36247	86954
06	31146	39118	59780
07	26297	69939	95129
08	79425	44606	64167
09	11258	07626	42922
10	57049	98540	20473
11	35889	62263	93680
12	93352	91232	70151
13	48138	90824	43366
14	10024	79093	10955
15	11418	19683	44046
16	57197	89115	68772
17	46181	16293	92931
18	88209	25430	17633
19	02866	24545	43027
20	96318	10874	67010
21	87072	41143	85784
22	79383	02455	58101
23	48014	14341	08849
24	18570	39244	92277
25	16262	20978	99923
26	65761	16877	28735
27	06883	65447	13075
28	59954	09337	08531
29	37116	08247	03407
30	58114	30488	77019
31	58921	27804	40264
32	25235	76317	85374
33	65788	34979	32086
34	31041	80146	17808
35	82178	97246	92373
36	12689	55313	99549

	A	B	C
01	69435	03862	47504
02	93577	74610	68832
03	50765	33996	55265
04	93132	76809	82017
05	97594	07871	46841
06	25690	95298	96485
07	81500	48217	44106
08	25942	20004	67720
09	18863	15876	98711
10	74901	29745	60896
11	81671	91817	47135
12	71992	73192	69440
13	50242	31330	77883
14	84020	25161	94931
15	48411	37366	48234
16	15634	71222	02407
17	34819	35071	10670
18	00087	03527	47700
19	35598	42855	13803
20	86231	91479	43617
21	65176	87721	95531
22	05001	73254	24268
23	94489	84821	51488
24	28405	83703	29281
25	80740	68317	75114
26	36556	04310	63858
27	66854	67662	64016
28	75000	92139	75266
29	46089	73057	40916
30	71967	62917	22990
31	76499	75908	51992
32	72727	70277	07933
33	86478	84804	18866
34	15506	46658	65808
35	06469	14262	97919
36	96017	15891	63004

	A	B	C
01	30073	30173	55128
02	79227	53511	52694
03	88178	16341	16992
04	63748	00687	44720
05	97639	31896	80084
06	49239	47864	36903
07	15381	95543	79402
08	48080	51297	76761
09	37812	04727	56872
10	31708	01424	13312
11	86222	46719	08181
12	12380	56416	01281
13	24424	42755	08532
14	77968	92943	37166
15	20540	76516	53764
16	52278	41379	23869
17	00045	72448	26732
18	85713	45795	96914
19	17079	87253	64718
20	45641	42889	71187
21	58932	70300	10142
22	18890	18849	69723
23	26973	32099	72863
24	68074	78948	40771
25	86674	34099	81561
26	07815	36861	94290
27	84173	07484	00506
28	97831	88813	04316
29	83469	24997	24410
30	10493	49385	77333
31	05110	50305	80295
32	35732	86000	71470
33	53866	20796	43778
34	97692	53318	68699
35	70308	34357	43682
36	81109	57028	65579

	A	B	C
01	73052	91050	38209
02	97289	00368	16679
03	14031	59978	41165
04	23870	32405	43753
05	52356	37825	25027
06	46489	93562	25595
07	73909	29595	22454
08	87974	82671	24969
09	20206	81584	10078
10	80897	24527	32617
11	97053	82463	19903
12	61535	38216	67619
13	13797	72122	63763
14	84678	72874	89868
15	98305	35665	29538
16	58845	11469	97718
17	27361	06885	41255
18	81247	45732	67003
19	59295	99222	79252
20	70122	19172	30566
21	51510	79013	43507
22	21208	20028	39121
23	03589	41118	43867
24	07255	95039	92042
25	11297	56140	66017
26	23986	26776	42509
27	88050	27729	24914
28	39881	71817	59420
29	46915	46302	50620
30	63649	42758	44860
31	32643	38839	73119
32	70654	21768	83639
33	80998	82525	15727
34	02483	23617	25869
35	60003	91911	39276
36	54959	71593	37877

	A	B	C
01	69683	77886	43181
02	49916	12919	14035
03	87856	81138	09014
04	71295	61340	15192
05	59304	86086	94861
06	17754	06227	87416
07	84754	39712	27457
08	87707	38403	69120
09	77902	56606	28133
10	93286	87866	72000
11	33437	51520	20229
12	36233	64594	12956
13	43667	28194	49775
14	25592	75666	04035
15	35202	89119	10871
16	03366	66487	14228
17	57667	64382	44160
18	88007	73848	91070
19	62626	23044	93380
20	56177	25028	12859
21	90981	77556	65467
22	74391	77235	90922
23	78776	85058	85657
24	66315	35049	42945
25	20690	07563	56979
26	68509	20662	21596
27	97364	92106	96045
28	70002	96991	70757
29	05334	65100	82213
30	02908	31919	84844
31	44708	51026	26224
32	05218	05209	07011
33	40719	15960	04885
34	67730	13566	43340
35	57548	29081	74637
36	42805	78758	54036

	A	B	C
01	14320	08425	12052
02	97378	34194	00415
03	45924	31358	55477
04	13532	18119	91513
05	44740	06881	34977
06	61480	90016	51309
07	74168	41411	45251
08	68364	87834	95827
09	98556	05773	12103
10	21125	52181	83835
11	37005	62022	39757
12	61607	83379	96841
13	30915	58460	09874
14	33167	57942	79251
15	50234	95327	79439
16	91302	38519	55421
17	43518	30558	16170
18	25119	70478	66420
19	59854	53164	24482
20	72349	89941	27462
21	36293	26868	04677
22	04824	29267	61547
23	82588	20890	95694
24	54434	17400	40729
25	74824	85532	20082
26	45848	87958	33788
27	23919	27742	49120
28	32142	70973	73892
29	06187	44600	95556
30	99880	42211	72982
31	32130	82779	12850
32	10292	44480	27835
33	35595	29905	78339
34	66354	71025	33329
35	30546	36648	40805
36	52636	74862	25167

	A	B	C
01	44536	78604	62529
02	73954	00642	55518
03	31973	42499	72743
04	44993	64632	93718
05	76091	59383	68756
06	56393	68031	89662
07	83868	95563	56087
08	86317	33881	31699
09	05124	51898	65081
10	83801	08319	39220
11	75184	64532	45244
12	58965	95265	44779
13	89614	15093	58120
14	47685	57591	62284
15	29484	90780	26764
16	68666	86069	30275
17	14265	03462	88228
18	57033	39698	33270
19	94277	49063	66074
20	00108	09479	15741
21	38100	62691	98329
22	60476	61847	53723
23	68384	19608	00172
24	04536	02014	83984
25	70965	27366	82311
26	84660	98832	05250
27	35611	03449	48064
28	77016	55069	77166
29	22271	21274	28542
30	27932	74986	25423
31	70566	71218	05204
32	12233	93007	96889
33	52388	43022	19055
34	96046	88217	37081
35	99044	90748	37829
36	69971	46726	97074

	A	B	C
01	34614	91629	65313
02	25743	61386	43232
03	59997	26317	53159
04	60878	33684	69908
05	81957	77823	05497
06	48397	35320	26928
07	43047	56404	06170
08	06654	82220	18296
09	16504	42557	13866
10	05723	67892	77291
11	51724	98573	50177
12	14572	42871	87549
13	04164	63892	29917
14	06482	20360	58163
15	21652	35350	57077
16	52838	42311	73806
17	04687	83563	48564
18	61434	63034	28242
19	32587	16448	41935
20	97281	12081	71331
21	79271	36131	26818
22	30613	65926	77836
23	04716	29427	33636
24	31265	32499	64064
25	22748	22588	26640
26	04191	81288	22521
27	60246	86793	99993
28	24370	51090	13409
29	99083	91712	44103
30	67484	58996	80106
31	79140	37584	81775
32	72151	13262	78789
33	54721	72269	42438
34	05267	75184	75404
35	10628	74169	11990
36	28478	89013	04312

	A	B	C
01	67887	01640	47490
02	54349	79992	03519
03	43712	31750	02356
04	07188	27715	09925
05	25007	63350	50320
06	46628	27211	60342
07	12675	20239	93697
08	35682	71526	77276
09	22767	79517	00980
10	97505	06766	49829
11	21198	16175	95271
12	14789	15372	28823
13	57734	99761	92061
14	24588	34372	29356
15	44133	89758	46509
16	43795	85892	40439
17	66918	73932	04968
18	25384	49831	54783
19	42496	44142	83291
20	43137	79680	63574
21	84455	11142	44594
22	24423	98173	10213
23	88387	03420	00534
24	98859	58235	38883
25	54895	56465	80885
26	64011	42057	14895
27	88258	11146	44528
28	16275	08387	90342
29	56042	28747	73251
30	96387	92873	01754
31	68369	41713	66277
32	53755	65463	60817
33	38418	22933	84103
34	64694	36220	94831
35	04570	11601	46142
36	60726	88088	63836

	A	B	C
01	81362	39486	08226
02	82942	41683	59850
03	44597	21538	13670
04	13160	64444	35970
05	65202	20779	62044
06	12930	12053	48629
07	48591	92667	18680
08	80490	80947	36073
09	42882	33278	45368
10	03436	55249	20077
11	17220	87535	40783
12	90604	18464	31128
13	86721	11329	88252
14	06102	23659	91894
15	04510	35259	17596
16	24253	14541	72119
17	95885	62102	62318
18	39072	83373	71822
19	66172	27041	65675
20	83646	13774	94164
21	02778	32775	62365
22	62740	19614	54863
23	76970	41032	25311
24	18491	71187	08783
25	11976	71748	65090
26	11424	65854	27273
27	76831	64402	54833
28	86639	55327	86963
29	84391	32523	01690
30	90434	96049	70495
31	78177	45596	04605
32	70844	28487	24269
33	14921	39965	37979
34	75442	80766	06640
35	50882	17089	82810
36	19331	21916	52350

	A	B	C
01	40667	60477	58955
02	20008	19099	69960
03	20976	31848	27627
04	26326	23181	16886
05	60775	36593	43044
06	87672	51577	28006
07	59806	79977	36006
08	33426	23292	43117
09	32540	87850	80058
10	68831	77971	12717
11	28477	20415	65913
12	07601	86807	15048
13	48457	60686	70555
14	26615	78432	41574
15	51453	83972	50639
16	09550	87023	46899
17	66539	20357	65035
18	78116	65169	65673
19	24991	74066	65015
20	17423	43508	01223
21	55045	75157	73788
22	05784	89170	78165
23	94817	80474	27644
24	35496	42823	34638
25	11676	31226	89266
26	84937	33055	61571
27	91370	01732	91247
28	45219	69487	41479
29	69734	97429	54874
30	33299	54518	06137
31	39351	81738	76127
32	13388	60781	87940
33	42442	46333	09895
34	16220	08257	75960
35	23869	07334	18320
36	17829	83658	71078

	A	B	C
01	99005	30859	89929
02	60104	29274	52151
03	65670	05873	32206
04	16623	58283	77699
05	82156	81220	05330
06	50528	76900	04832
07	88117	04673	67873
08	60044	67204	24217
09	24685	33081	26606
10	92234	94760	04808
11	94451	62376	05981
12	39361	36312	52764
13	04537	59969	38837
14	59860	16425	07908
15	08635	80571	62001
16	82173	57016	42597
17	92365	86781	06990
18	11984	41116	51860
19	41682	59139	73887
20	48918	34706	14590
21	47695	12534	73294
22	44092	67723	31705
23	57517	82010	83643
24	19095	18794	70726
25	19983	63444	49403
26	14545	06771	48098
27	26138	39207	80702
28	85752	52710	22748
29	44411	80588	05114
30	17566	51817	26170
31	57753	30755	61962
32	78715	99179	79025
33	09111	51244	75698
34	62426	16059	28604
35	36051	20623	57240
36	30689	37691	70936

	A	B	C
01	73465	86602	61882
02	38921	78540	71218
03	96870	37635	04102
04	97215	57564	41727
05	32635	75856	53312
06	90098	78009	01743
07	34906	96296	35723
08	39686	62081	10458
09	34060	05616	68493
10	81526	99788	35139
11	31953	36862	79104
12	13831	46395	48465
13	24943	81338	08122
14	12491	03805	92325
15	86254	03656	70352
16	08860	84649	02917
17	53235	42156	02363
18	62921	54023	46160
19	41141	96691	52870
20	04618	86561	83010
21	94611	42253	25789
22	48758	25817	44092
23	97048	92324	29352
24	26150	32649	27767
25	37232	60101	95024
26	48201	81214	87577
27	93461	17868	40979
28	74330	05858	71815
29	64294	69721	31532
30	41004	48190	95770
31	71882	54282	72945
32	05199	60790	48116
33	02655	57505	00780
34	82356	38679	03309
35	77064	53289	35356
36	87962	25021	68222

	A	B	C
01	11821	89189	42475
02	15238	05593	58649
03	74194	54098	71889
04	30700	02183	08006
05	50638	46038	18088
06	03798	78352	29640
07	76779	78260	44795
08	24840	11165	62395
09	31438	74981	71131
10	96165	57919	86568
11	43234	09537	25351
12	56273	74624	20860
13	45615	00684	75999
14	84900	97745	84030
15	63839	39799	17651
16	24669	51256	07759
17	13950	43109	88026
18	36256	78397	45598
19	30773	21080	83760
20	21195	20865	52339
21	57532	11187	17149
22	69293	29604	84852
23	01068	20125	92656
24	88813	36330	50519
25	83240	97056	99623
26	01585	78598	80637
27	40337	39404	73754
28	44958	06984	85372
29	74622	68977	68568
30	76855	00513	92527
31	08934	87103	27706
32	10723	87326	16004
33	15780	29674	68979
34	05013	12249	90800
35	94126	74673	92513
36	19582	31814	96479

	A	B	C
01	47679	25855	67530
02	27995	15623	92506
03	93771	91668	58981
04	22066	41671	07042
05	06970	43914	06354
06	76142	72648	56739
07	70705	56726	89175
08	47189	13591	77005
09	83511	51155	45924
10	26356	17299	36688
11	31329	20475	53271
12	19170	77263	90131
13	75043	66719	19894
14	03606	67878	00103
15	62910	13259	45409
16	05069	61094	05087
17	18946	98460	37914
18	98541	81318	97725
19	90361	92219	82105
20	00030	44587	05189
21	42565	40657	03515
22	96372	19884	24083
23	13138	61636	77680
24	28352	33504	97500
25	78331	76098	68537
26	77486	60455	15142
27	58726	76371	46280
28	30782	80942	98159
29	21986	00441	63609
30	39950	70193	21842
31	11425	35676	80943
32	59958	49394	56575
33	22755	22816	61030
34	16039	03128	02224
35	22565	59839	26562
36	31596	43485	01539

	A	B	C
01	49610	62065	45556
02	20520	49474	48894
03	27716	62724	58146
04	08432	11579	10266
05	49867	34798	23148
06	76364	08583	12919
07	64444	76833	70234
08	70616	80043	61728
09	74462	26173	24135
10	84692	28203	55764
11	54193	16802	22687
12	81079	95241	68736
13	52796	97763	16921
14	11937	76796	84135
15	82274	08430	67967
16	77626	23271	05717
17	32911	20841	82477
18	48610	23285	92460
19	56970	94509	18232
20	00736	34750	50143
21	48787	17121	52735
22	65359	67907	01437
23	70332	31043	81802
24	45989	02874	72707
25	80021	34325	52702
26	89026	40045	97902
27	78246	99294	10998
28	40178	20893	08247
29	24936	79643	60443
30	28293	18774	05225
31	10389	39149	34288
32	23330	99500	85597
33	81226	84309	17349
34	33309	55001	86346
35	44542	24551	69367
36	08457	18353	23909

	A	B	C
01	57316	31515	27922
02	13305	55311	15033
03	05692	45506	07023
04	21450	62107	77530
05	07918	35869	49252
06	66761	34711	22534
07	23246	89133	26324
08	32281	91084	04963
09	37200	49623	46130
10	97354	07165	65870
11	67934	06359	86173
12	72473	64814	24965
13	47312	28943	20294
14	55084	46806	42531
15	16963	55168	63097
16	35445	17958	13782
17	29277	59843	60175
18	91980	75330	46837
19	44237	10509	76615
20	54783	81545	13501
21	03357	72526	21760
22	62707	38160	03222
23	85986	00928	70284
24	00417	40745	48045
25	45829	68560	85404
26	84249	58652	11164
27	30478	48250	54026
28	62582	37523	96541
29	97634	87038	36656
30	93126	43694	52230
31	28083	39577	28378
32	88600	33162	34907
33	23225	69989	66630
34	31649	11765	35581
35	74910	08362	64883
36	25607	12003	22404

	A	B	C
01	84903	21167	96958
02	73809	59894	83698
03	41425	88665	93115
04	44912	21607	89160
05	51352	64008	07012
06	42301	65469	00950
07	18935	20124	31853
08	97257	76187	42836
09	61814	19957	58982
10	15567	19271	24952
11	33629	38763	22842
12	98714	91138	34919
13	22570	76375	17505
14	27395	62214	52139
15	74435	55042	82494
16	58334	55548	20873
17	60097	35449	75010
18	29552	23347	35765
19	21237	38187	51920
20	98842	75323	71395
21	82530	48859	76415
22	38437	66778	32495
23	59239	77584	24058
24	82094	55677	64910
25	18137	48316	30165
26	49902	13296	15286
27	26538	01946	56874
28	67475	91951	80865
29	41731	69221	52977
30	60064	16803	75960
31	34707	57459	62345
32	52882	36163	58074
33	83121	01301	23656
34	71633	09732	58200
35	59853	13851	77594
36	61124	66462	64565

	A	B	C
01	26539	77299	39233
02	88334	60683	20247
03	23062	83572	70776
04	08260	18875	18379
05	74507	92217	12111
06	05900	07067	83321
07	23399	56239	25966
08	97511	26564	88324
09	50028	67070	75352
10	88555	61634	07272
11	20431	29905	71368
12	47294	39219	34614
13	57950	14466	31644
14	43179	83939	44132
15	00102	90962	97610
16	70046	95073	29405
17	71298	73650	84527
18	25218	52245	47455
19	12904	45799	71421
20	03127	21593	15115
21	79401	53723	45558
22	06126	58076	54149
23	11704	98226	79914
24	10024	50713	40813
25	32446	90573	77114
26	40696	94292	56625
27	03763	68032	26406
28	66391	27538	43727
29	66887	56934	46906
30	74852	92592	35566
31	02542	39920	24169
32	60182	78286	91866
33	98865	95977	78479
34	64749	03412	50281
35	26902	18887	33502
36	17293	89264	49211

	A	B	C
01	64928	72803	53865
02	03674	46705	51216
03	62007	92149	36364
04	29260	29541	92296
05	83727	49980	61186
06	77826	77374	12276
07	11795	82821	97318
08	16570	39389	38515
09	04823	22143	76957
10	66424	38822	64959
11	19614	39276	59586
12	99402	25239	93293
13	25418	13357	51823
14	55731	21640	43898
15	25995	76280	52817
16	23638	50675	56434
17	29859	81110	10420
18	46651	25328	61484
19	02075	71718	43262
20	41766	68552	59935
21	55697	19137	19138
22	88899	49478	45217
23	02933	74844	03065
24	48194	10064	75732
25	24606	89683	82157
26	62775	35762	27817
27	44266	75497	32344
28	09537	85371	01844
29	54103	48676	96665
30	94616	93120	79835
31	75775	15716	36440
32	76125	67786	09641
33	20715	43040	59733
34	05440	74654	33415
35	55633	66100	72160
36	98101	95999	82959

	A	B	C
01	98514	59238	26941
02	21029	78115	91790
03	15282	05764	81373
04	46186	83895	68437
05	16182	00637	47451
06	02247	50525	82575
07	29343	06327	65070
08	93821	15837	38660
09	89964	56860	43495
10	34521	02060	51841
11	65145	07302	26758
12	92980	49406	33223
13	89515	89823	08610
14	46418	82592	16677
15	19521	90916	21984
16	16625	51089	60714
17	75819	36685	11021
18	31545	77468	91015
19	78671	39485	40193
20	76646	79498	47003
21	98220	73795	81712
22	12469	14805	74509
23	59363	41990	34106
24	99115	14413	39255
25	92883	05970	35636
26	42250	82649	50005
27	04265	95621	61980
28	20018	41716	82260
29	45853	44257	42582
30	58811	23615	21677
31	44649	63699	32261
32	68415	31485	62428
33	71537	49521	50694
34	78298	60709	69517
35	94138	42078	30188
36	49347	77588	21525

	A	B	C
01	32987	00959	87253
02	92626	14171	37473
03	07260	67826	38536
04	03719	82406	14273
05	35146	30370	78323
06	38905	79345	93503
07	62787	78183	26689
08	50813	83298	54741
09	44233	13148	72648
10	23834	34185	02309
11	22958	68981	93663
12	76079	83714	43642
13	23027	44177	60915
14	79341	16040	54869
15	78278	10897	16121
16	65277	37891	08494
17	96964	77573	17163
18	25114	32601	33855
19	21356	68245	31490
20	01607	40290	00812
21	81153	02960	00470
22	50301	90307	49532
23	00889	23294	86445
24	31056	81177	76650
25	67541	35668	26352
26	41516	95912	41637
27	86343	40020	91434
28	56438	83819	14226
29	88857	10690	28112
30	02647	52464	21493
31	03898	50186	16332
32	16968	99934	42151
33	50957	45116	34837
34	19392	37567	63911
35	35547	66869	57995
36	09212	99279	15925

	A	B	C
01	38530	98075	48544
02	73082	34013	51536
03	09597	95975	99863
04	86946	59620	32079
05	28744	28959	13648
06	57878	42041	66557
07	98203	17368	70335
08	22200	83353	78675
09	67627	97149	97042
10	28667	61554	96594
11	85518	97984	93098
12	33161	90974	67984
13	23015	93877	70767
14	15205	95819	21314
15	46756	39398	89483
16	77198	72002	64601
17	91250	92968	44267
18	50553	00462	04594
19	24262	73876	00433
20	86153	64523	59952
21	26895	71537	14097
22	42590	69331	81539
23	61051	80168	97651
24	25875	19602	45188
25	63285	04487	32459
26	87866	19844	24134
27	40645	81749	01011
28	82658	76097	84887
29	98547	60653	81869
30	73752	07367	80097
31	24229	72555	38848
32	32035	90198	92852
33	36918	99632	13177
34	67437	80559	93358
35	85691	61707	02113
36	60559	03086	98435

	A	B	C
01	42181	43450	65497
02	98017	81004	97816
03	36012	52385	26744
04	35860	32699	75457
05	12907	95922	72017
06	14818	24810	49261
07	41224	41505	89897
08	09395	03883	68352
09	12856	36256	41933
10	81479	52681	25979
11	73917	55939	78048
12	92770	99029	66875
13	64507	50901	99023
14	77906	36892	88977
15	12716	78378	46328
16	34627	35015	49214
17	48267	33156	11600
18	03833	47180	91791
19	19931	20811	20491
20	24854	18449	07651
21	22866	19690	48877
22	70298	27343	40070
23	52588	72949	56246
24	55328	03883	43978
25	70740	37698	96006
26	44994	18062	13928
27	60085	53144	59219
28	97696	12072	79419
29	86801	31944	23582
30	63677	06169	01067
31	44129	29095	02576
32	49513	16925	80526
33	60593	42519	57629
34	80972	43780	72682
35	02620	74899	36273
36	36258	97459	70671

	A	B	C
01	88546	93375	59518
02	16817	20987	91164
03	49812	06492	84717
04	37656	75574	78957
05	09802	79132	10943
06	01754	69612	02909
07	35314	94980	70986
08	60863	45415	57018
09	64922	78337	43009
10	53510	65568	43348
11	75860	39051	61564
12	29694	22262	74697
13	62683	02521	18915
14	01263	56589	51498
15	27043	62529	77881
16	61211	49824	14095
17	61667	95044	30319
18	28967	17209	99325
19	59358	53873	85805
20	15409	97793	68899
21	25210	50137	97334
22	03399	95891	59978
23	22706	62647	92618
24	56991	66446	36043
25	63191	75024	94318
26	62901	98529	15381
27	89335	87816	31840
28	92498	90571	88909
29	47505	84053	32951
30	08192	96088	44358
31	30918	22188	35402
32	79865	69537	43195
33	60793	03322	79997
34	49393	07641	07764
35	12221	59053	51995
36	36697	97737	50831

	A	B	C
01	69626	27198	54117
02	64268	08812	63071
03	06344	23277	24636
04	46757	98570	82638
05	43026	32246	67659
06	03376	71602	99382
07	85977	49829	99507
08	32020	00336	68711
09	97446	78609	20969
10	41349	56149	46918
11	14564	29369	57130
12	47529	14539	17241
13	97305	44991	74594
14	62908	57266	11103
15	19024	94879	12753
16	83103	33777	58775
17	18310	52848	44682
18	05511	87514	93573
19	11772	04598	86955
20	72317	31101	83465
21	96831	28824	57367
22	23598	93578	69646
23	00245	34527	83058
24	55350	55917	52261
25	08491	80861	04105
26	64334	23604	67211
27	59245	75752	66201
28	98899	57364	20512
29	27702	07650	61029
30	62040	13478	54179
31	83897	57898	26525
32	57617	55501	30333
33	31965	44492	15723
34	70695	23223	84134
35	03707	79575	91553
36	47883	65938	25722

	A	B	C
01	51474	16965	44940
02	66428	31735	95202
03	85297	62962	38659
04	51198	53750	10400
05	86733	46234	18095
06	53935	18639	71594
07	91825	29627	54653
08	44564	07536	56606
09	78434	76442	00640
10	66422	87745	78610
11	67078	82030	12676
12	74783	49003	51348
13	27286	36689	90899
14	54312	91859	24894
15	98091	45167	58546
16	93944	33843	63222
17	84056	85036	03386
18	22168	91040	21787
19	47228	79514	13888
20	18575	82456	96161
21	44052	31389	44204
22	65746	65365	76759
23	49671	92647	24754
24	70219	65251	77408
25	36101	85276	06496
26	74951	73742	92915
27	84592	82455	69417
28	42140	67020	49968
29	98710	55427	02955
30	37731	72174	55208
31	88900	80501	40168
32	11339	79970	96793
33	84048	84785	54342
34	68906	71718	49148
35	72196	27913	26471
36	39970	68586	79433

	A	B	C
01	41326	61048	86295
02	93578	48452	40831
03	45460	51301	45978
04	09129	62548	28074
05	02274	76634	37302
06	78672	95337	44583
07	91682	07812	95234
08	71841	85624	64093
09	05207	14496	52897
10	13574	44744	08579
11	07734	60208	08100
12	77839	63295	26293
13	29022	04494	99508
14	86254	50099	31949
15	28095	78292	48119
16	92827	66524	62245
17	81333	74092	01757
18	79247	67182	80216
19	81955	36648	37426
20	61603	71099	57317
21	39473	25493	50351
22	28366	54410	54889
23	89954	78210	87907
24	51669	32339	30101
25	92036	12726	20110
26	43266	68111	54928
27	82169	74136	66885
28	06682	13378	74373
29	53062	02723	50733
30	66550	20831	02846
31	48563	36744	67935
32	05867	56110	00681
33	24809	12296	27035
34	46017	74953	56400
35	01977	32724	17661
36	12458	73422	20568

	A	B	C
01	71470	66633	22788
02	81001	78717	95318
03	29486	41182	97370
04	58383	01217	05308
05	40966	07713	14166
06	13652	75843	45836
07	65420	85378	02898
08	37620	73188	61740
09	71887	22300	11759
10	03327	66258	38777
11	06446	74564	93857
12	02969	38562	69829
13	93240	76227	89806
14	88810	73495	20939
15	76852	86753	92577
16	26847	01562	23514
17	88931	66146	20082
18	48173	98618	61281
19	53267	90131	38256
20	45429	99047	94385
21	27695	70812	59836
22	75986	88820	52609
23	96249	10715	91081
24	48454	32556	84551
25	44648	71171	90057
26	68197	61631	34031
27	22256	80668	53414
28	47597	61450	28164
29	84667	86506	40944
30	00270	87280	80639
31	33380	99145	52347
32	97589	80592	18648
33	46000	34081	66945
34	02818	61716	27941
35	08142	37967	89450
36	83891	79225	56942

	A	B	C
01	76807	03790	23303
02	75201	60785	58359
03	91745	39792	28459
04	35059	25036	67122
05	32004	96953	23804
06	56630	56691	59544
07	16159	94927	64828
08	61781	35083	71234
09	79856	50356	84130
10	71511	70216	60824
11	36257	87894	26323
12	15088	18974	30198
13	23710	67914	22959
14	14536	28642	89786
15	13756	85214	31553
16	44943	36588	28411
17	72362	13349	11954
18	63123	22785	70202
19	62284	35957	97939
20	15138	86233	89169
21	83228	05060	91181
22	28226	69252	08479
23	01399	57212	15407
24	43207	96855	33383
25	79957	90134	42248
26	04426	35117	25682
27	74801	91824	55043
28	90817	67316	35018
29	04452	27920	38898
30	10976	14039	26115
31	37909	48655	38408
32	55877	70000	04760
33	57763	75906	44200
34	71444	66668	32042
35	71929	15230	70458
36	92875	73902	15989

	A	B	C
01	11379	31164	75454
02	46903	76857	48865
03	37710	85450	73876
04	70877	50115	96675
05	80076	55987	03229
06	91052	17587	60515
07	14896	81302	29526
08	91507	67072	76740
09	57461	47251	81836
10	91005	33182	96123
11	49797	13664	29820
12	28089	39697	87432
13	83114	20818	91256
14	78343	77865	08700
15	94473	49988	68442
16	58951	25384	47546
17	04678	46772	90652
18	52229	63260	75295
19	79181	10595	56328
20	34891	71318	80680
21	27484	24772	90286
22	91018	82367	16703
23	35739	55039	21323
24	02095	93999	40318
25	43087	91610	21722
26	26227	77278	19092
27	43169	06653	11471
28	46212	65699	67171
29	19010	41353	15269
30	55430	26885	54837
31	08450	68784	76671
32	96883	08093	91513
33	33070	26012	97838
34	10034	21232	80316
35	98030	25213	97036
36	43105	43026	43530

	A	B	C
01	58432	91125	85448
02	65668	24637	38965
03	78633	75086	74123
04	12804	95296	32719
05	86637	98613	18774
06	07539	82035	79209
07	93483	06446	55970
08	61203	27022	88669
09	48626	67677	94149
10	39111	53359	83146
11	40929	10257	21762
12	87776	46488	03159
13	10438	26564	36300
14	82449	25056	18649
15	12565	25415	77808
16	30076	53277	72870
17	67075	53864	66287
18	62867	91951	99903
19	58681	09938	50881
20	62913	26653	59583
21	63918	46331	60326
22	02104	84185	95688
23	56918	61819	42912
24	86272	47385	79283
25	62384	30542	31552
26	31097	58097	63250
27	45210	63136	84739
28	31487	25378	48154
29	70635	70066	79711
30	70660	15569	40601
31	92855	90998	85794
32	25171	65233	41847
33	31594	54243	17257
34	52130	75771	81420
35	83073	53357	97909
36	39564	69837	37629

	A	B	C
01	10224	44881	63323
02	06536	65016	97277
03	86756	28529	45576
04	78570	61396	52714
05	73730	22451	89981
06	88460	99517	04608
07	27169	37220	63296
08	61618	86310	50672
09	98128	53700	11781
10	96248	36065	41783
11	72580	53154	48733
12	08032	33084	45610
13	98265	64634	07386
14	49447	54131	76910
15	53043	82546	79642
16	15309	24521	29804
17	28906	47679	95245
18	55873	92579	14367
19	10278	51683	48812
20	18251	33471	42369
21	80288	76805	64528
22	08835	20849	65949
23	06510	83653	55488
24	46413	18861	44840
25	03464	52924	34365
26	67207	00772	16631
27	41632	48701	53145
28	62868	49881	59120
29	42430	61366	17648
30	34907	19699	01632
31	70725	45928	71392
32	01274	01147	79608
33	04034	49455	91197
34	45066	61168	92363
35	41886	40999	26169
36	94850	36300	60143

	A	B	C
01	22949	18332	92407
02	63969	10539	08540
03	10880	26666	34408
04	26552	53526	97401
05	20222	61377	24962
06	17930	09161	00545
07	88374	29693	25705
08	18491	69247	20917
09	14868	63543	61669
10	67405	64152	76025
11	92216	42814	30756
12	00562	33820	85279
13	23060	93664	91887
14	70217	43596	27046
15	77697	74113	31504
16	46940	05560	77298
17	24087	99326	83179
18	50591	64346	85079
19	16645	38334	36569
20	29680	38443	19620
21	16204	39983	81484
22	73611	11095	25510
23	03225	49900	73582
24	76257	79403	57182
25	83525	39380	54089
26	89910	42253	68668
27	74151	01091	74660
28	93482	32209	28653
29	59078	23162	09310
30	18241	35790	71811
31	59819	68823	13442
32	33687	74070	40466
33	91335	19076	40592
34	72411	76665	30328
35	81826	75937	31909
36	44895	28176	80313

	A	B	C
01	32221	33747	18559
02	69347	99942	31888
03	92889	35473	65909
04	44015	39262	43631
05	30936	22630	81275
06	82409	23927	72836
07	74987	88073	06930
08	75794	87557	58882
09	42409	23437	15376
10	87106	16929	92021
11	76342	75757	17764
12	22281	21567	53803
13	96410	23743	52732
14	20141	38870	04738
15	43106	40203	96903
16	99612	89334	84517
17	40003	67865	98767
18	89422	51586	55738
19	89020	10096	52832
20	58797	00100	88231
21	47722	93317	17174
22	68312	63597	05255
23	20646	07919	81544
24	84962	12079	74955
25	72277	53260	52190
26	38435	02963	35517
27	45495	33591	02030
28	72224	46646	04543
29	23065	56946	95257
30	20262	05126	90990
31	79705	18549	59906
32	08740	43972	25513
33	34226	23824	79737
34	34689	55575	25447
35	20632	15915	66745
36	80273	82591	23747

	A	B	C
01	62429	35261	07033
02	36761	73880	93061
03	36640	71081	21235
04	01667	52891	80398
05	11810	00464	46714
06	66779	01514	87655
07	00756	78021	87708
08	07841	79710	71535
09	41107	39123	65829
10	96073	76345	31548
11	09187	05055	56523
12	78090	25027	33166
13	78078	61708	82567
14	90611	04553	55090
15	71756	48960	94536
16	60980	35896	90830
17	64084	59213	96527
18	54870	59917	99235
19	95210	92476	11715
20	51317	33725	69360
21	92078	62704	67216
22	10553	56667	08631
23	63494	12070	06266
24	76845	11190	89728
25	78398	16615	54814
26	75472	67202	99224
27	95869	74888	13622
28	97164	84544	18883
29	76693	81735	40012
30	38386	23116	97254
31	40183	46432	68855
32	15057	57856	02655
33	76581	65036	27822
34	64265	66624	40137
35	96522	01728	32297
36	43279	78625	49925

	A	B	C
01	94323	75902	08314
02	19185	73632	76099
03	81396	27682	99505
04	59960	68152	66609
05	54366	80851	12513
06	62482	30097	69547
07	75206	25526	66595
08	99731	97082	22661
09	69062	38079	98976
10	37798	59323	03588
11	58999	39959	14135
12	13694	87134	50676
13	13868	74548	21818
14	61388	01629	97913
15	98119	09763	17205
16	43698	47453	43781
17	38802	06782	50237
18	41320	75911	92814
19	35475	78932	18989
20	24120	69386	16789
21	51710	88725	84433
22	52967	60582	59278
23	24495	72272	50563
24	08758	74778	68033
25	81818	39795	77660
26	29969	63979	36773
27	37692	78876	37283
28	57491	62678	77538
29	54594	89517	57933
30	73821	75220	19334
31	24275	79009	05348
32	26949	41184	75702
33	59004	62659	88745
34	08002	16595	65340
35	42932	91363	66280
36	49646	36797	52618

	A	B	C
01	08136	16985	28730
02	19086	05584	91428
03	43483	71205	40932
04	61020	49985	18502
05	74636	35968	11986
06	52983	45362	43282
07	24341	57815	53333
08	05714	98220	58236
09	94441	83706	13720
10	94620	70845	46059
11	29468	64398	32734
12	58689	97073	95862
13	32381	89838	78965
14	21236	34329	31166
15	13993	37069	66108
16	05192	37532	17848
17	27916	19596	74707
18	40099	29317	44408
19	75749	16620	03551
20	14341	93544	60896
21	62797	68359	96736
22	33325	31884	45866
23	83777	13808	42020
24	54941	50055	58804
25	07982	24337	60443
26	12623	70490	87077
27	44952	96503	97315
28	98447	68233	86372
29	20748	69197	24857
30	87788	53330	51801
31	54200	03282	38779
32	86326	33608	33919
33	58088	17478	72838
34	13591	19523	44734
35	77654	66270	51315
36	96892	55702	53349

	A	B	C
01	81616	17197	94582
02	19693	81745	83528
03	18812	49468	18487
04	83760	83491	77467
05	62703	68304	18659
06	47158	31081	75194
07	00958	01737	80046
08	31143	94182	13290
09	32195	44256	08711
10	74589	18849	52763
11	31100	83406	12185
12	96887	94069	14211
13	28003	27645	49655
14	51973	62374	60166
15	71083	22054	71688
16	16614	07519	37339
17	13885	39528	71782
18	01631	62747	74410
19	64717	16691	72611
20	41406	29218	55391
21	78385	36827	67864
22	53884	81781	19726
23	75204	22761	52044
24	14010	13987	16117
25	34642	11216	57093
26	48941	78642	22775
27	33980	84299	01451
28	23356	61521	91590
29	72404	14976	50255
30	96889	20180	06178
31	75071	95920	78396
32	60940	08793	54499
33	81724	24481	90347
34	70227	33553	20958
35	56436	40086	17527
36	15823	09597	38166

	A	B	C
01	54190	27092	31450
02	45843	72303	90906
03	03044	21839	34952
04	55451	56230	74990
05	85224	41002	36135
06	92397	14752	94199
07	79972	68501	19909
08	63874	49179	22961
09	50104	46165	75157
10	35589	34221	48661
11	96229	63562	57296
12	53654	30338	27386
13	44353	89989	19056
14	77240	55521	11876
15	29913	47779	47972
16	99871	94753	19250
17	20324	88055	09145
18	59090	21583	29116
19	59830	08705	16843
20	96265	70067	83482
21	37061	35171	26536
22	88582	99334	51371
23	45874	12756	75842
24	77299	27902	37043
25	82146	22646	79526
26	66224	40113	27556
27	58483	62271	10454
28	26885	61249	74276
29	43892	82287	21158
30	02440	53223	04593
31	12444	61886	92202
32	70649	08932	72461
33	04674	26003	66664
34	54482	63063	43937
35	44532	39447	36478
36	96853	62605	15055

	A	B	C
01	29181	96808	02786
02	16409	06712	51890
03	39041	17280	58204
04	94580	21060	37679
05	23969	10700	87292
06	20408	07997	62382
07	16342	28330	77630
08	75761	66121	86026
09	57179	51360	25937
10	35015	16729	10936
11	95884	08012	43235
12	70834	04564	55384
13	48052	74476	84660
14	48958	87490	74824
15	99570	59054	46028
16	30186	67293	67870
17	74543	37414	08546
18	30169	69362	44100
19	02489	31912	91350
20	24399	59085	90290
21	55623	68785	59393
22	26705	56403	25140
23	43864	36319	69330
24	68657	36225	61840
25	20686	34335	08606
26	27553	63308	56003
27	98586	98260	38976
28	62197	54911	31065
29	82068	53872	17809
30	97379	28382	60559
31	64692	21620	51422
32	17588	23382	64931
33	43138	68138	25931
34	20115	09761	11092
35	82381	37300	02558
36	84889	18854	52555

	A	B	C
01	58058	91057	87147
02	58433	81289	34727
03	57476	73891	70475
04	43235	91054	50815
05	06494	22320	91332
06	93469	18881	31815
07	83261	32164	39136
08	11861	52767	13739
09	69829	76819	91981
10	69512	79996	89792
11	62312	73987	68253
12	52707	14090	90251
13	90492	04497	88786
14	88432	88831	60948
15	05677	29039	62472
16	47082	16136	61374
17	07368	83087	38081
18	08306	23572	66729
19	10029	70401	41334
20	02579	49212	84698
21	47442	72452	69352
22	91548	01677	65080
23	88781	77278	32844
24	06933	18213	75772
25	61804	53529	35590
26	81890	81773	53792
27	88079	43438	24447
28	02134	43019	67366
29	67092	37858	07228
30	79244	67781	45049
31	73588	48852	38972
32	52762	08750	44635
33	67027	54457	48073
34	03342	90857	07877
35	40489	48748	11408
36	34011	35811	01337

	A	B	C
01	65098	49467	42905
02	40595	74617	61820
03	36449	41417	13195
04	35781	57957	30090
05	91798	91002	45669
06	15675	69882	84533
07	89513	23651	88142
08	13841	64840	23280
09	44451	01722	14035
10	97352	99528	37187
11	68479	87307	83184
12	92903	94229	77991
13	12131	04010	58696
14	06883	46119	35465
15	78189	94838	10371
16	86783	08090	82413
17	92764	18976	23747
18	42370	00722	95731
19	73596	21541	25048
20	83465	99675	58843
21	39677	48642	16878
22	16833	61856	62426
23	74371	33926	44135
24	19736	98004	27515
25	14590	77018	99668
26	78462	58463	03380
27	68210	41854	59200
28	98596	67217	67469
29	65203	94731	33402
30	16277	23317	91184
31	81752	38825	26767
32	74680	12931	88030
33	27788	03766	08915
34	29626	38541	99537
35	36851	56214	51844
36	61958	35264	18490

	A	B	C
01	31953	44520	58212
02	79547	17754	53063
03	81194	72546	54892
04	39597	90499	86616
05	63835	67312	45926
06	20561	35690	52575
07	66672	17783	99713
08	81945	93821	03613
09	89291	60675	12507
10	47976	11127	63873
11	17531	56674	30222
12	55602	04136	81752
13	65129	88678	23572
14	23056	95868	04158
15	25935	19375	47899
16	18981	75822	38285
17	93963	67000	95092
18	19873	44542	65445
19	59173	50708	69392
20	48533	13877	12068
21	23211	51846	58747
22	96650	60021	68251
23	72763	32044	40609
24	03908	64585	94902
25	30632	71595	33751
26	14011	73620	00424
27	13605	89689	08527
28	54634	32544	20013
29	71021	10301	78819
30	93532	53744	53084
31	74749	15506	19906
32	16647	48456	71360
33	04649	88074	71098
34	45712	45832	60970
35	01632	08858	67621
36	52204	74579	70039

	A	B	C
01	72289	59127	05590
02	35508	99696	68697
03	62808	31660	58602
04	05804	60529	25660
05	82319	88785	45197
06	16172	93481	55601
07	99091	03278	51263
08	55745	12254	34331
09	91924	02015	88877
10	31025	51049	05251
11	88976	23758	42777
12	75577	10549	19263
13	54558	93433	90963
14	49338	93076	88146
15	82226	67556	71209
16	57988	17168	61769
17	39876	06079	83010
18	77982	78785	36236
19	32492	53931	19443
20	37319	70220	10166
21	73428	14597	72180
22	25466	64923	31026
23	17434	74032	60113
24	29841	09069	83138
25	35590	89756	12034
26	61341	60540	55449
27	65495	91467	68915
28	47405	99409	17864
29	95445	24940	82251
30	88981	60506	13988
31	20798	38780	53561
32	11713	98703	79553
33	08442	01305	61153
34	32798	27223	70341
35	28311	45885	43783
36	43456	04410	09033

	A	B	C
01	13152	15519	81951
02	56941	24459	29626
03	63604	89833	51035
04	37895	81991	02744
05	64846	68336	85444
06	63569	83988	55310
07	58742	65471	81468
08	94287	03830	84600
09	63495	47869	81904
10	77215	42154	29213
11	23045	72501	68301
12	43491	10095	26627
13	13228	29550	98548
14	88006	68615	88028
15	71920	13379	34419
16	06906	77108	68982
17	63079	96775	92774
18	88715	97470	98331
19	56965	83650	67709
20	20760	11273	24500
21	80683	86351	45152
22	21907	35747	84877
23	33832	28025	83878
24	86006	03395	99283
25	22512	86700	22646
26	89663	77546	04499
27	59455	01562	52002
28	16281	44276	10659
29	00049	46842	71048
30	57885	15217	75671
31	50110	81790	14269
32	68823	13735	32319
33	18532	35192	74605
34	09098	45889	87662
35	18818	65733	65626
36	89901	69156	34792

	A	B	C
01	36753	44071	81177
02	50975	44470	03751
03	50583	91699	08485
04	46080	91478	96949
05	72088	60809	15033
06	42096	73099	94072
07	79803	72507	19867
08	09533	01786	61318
09	47219	50458	01432
10	21899	84708	88666
11	85373	42875	77400
12	55483	07240	02432
13	91014	30439	11410
14	26266	38429	30591
15	82795	59057	44814
16	60357	83682	05282
17	94573	46147	78820
18	98991	56430	31375
19	22451	1211/	62802
20	36489	56706	60245
21	02716	18552	43237
22	47016	89612	21664
23	21365	17562	08290
24	61476	91300	67699
25	34136	95293	29189
26	83914	14993	32230
27	15568	11213	92684
28	15268	15192	25661
29	61822	18447	35382
30	03617	57581	49955
31	57890	33890	05907
32	83242	07817	24933
33	01972	86180	24253
34	08310	04569	75638
35	69147	05246	58877
36	04985	76742	71242

	A	B	C
01	72049	40846	06600
02	44355	24659	96544
03	42621	98901	44712
04	19049	77535	17624
05	32037	09266	21168
06	51265	94369	74675
07	27502	28293	83286
08	69004	92042	31828
09	30373	18882	96325
10	80035	08985	36861
11	86390	66563	26682
12	71759	34375	25909
13	15481	46411	60248
14	32301	07311	70180
15	76612	96466	64494
16	30259	86368	31358
17	10218	89754	69605
18	56655	16299	18547
19	38267	18732	52837
20	74939	75657	22250
21	22075	35551	44756
22	27603	13487	29928
23	12091	77682	63963
24	39977	49989	09434
25	83529	96451	71810
26	08703	41833	96482
27	95212	41424	66851
28	21204	01628	39736
29	63480	43735	17686
30	80894	89355	88157
31	04723	76160	42412
32	89062	47202	61305
33	85485	04556	38043
34	34869	66586	35024
35	19104	92698	80865
36	43585	15477	38346

	A	B	C
01	06478	55517	21892
02	86564	97055	78343
03	92599	72572	70676
04	44143	99545	71811
05	36976	77207	61660
06	56364	35711	18620
07	56064	65226	28954
08	42147	29056	78110
09	48264	37107	83926
10	39004	46352	43558
11	65550	41873	97001
12	57087	11796	33293
13	49170	25192	35131
14	79552	67069	56009
15	62282	60024	10906
16	06242	02185	78339
17	37027	44902	99561
18	53919	68793	03870
19	88456	09546	53973
20	84464	71470	71496
21	57985	87523	08083
22	25755	67452	32020
23	95371	72782	21828
24	93044	87737	21288
25	97979	69623	25668
26	29568	60537	59601
27	33682	88741	94141
28	54489	02455	51073
29	58247	91299	67490
30	63839	90170	04602
31	21379	40329	85638
32	33667	70770	09448
33	29345	16316	21968
34	87221	33076	38609
35	08975	95009	13405
36	75982	60671	77615

	A	B	C
01	84418	04686	71680
02	14385	30218	29983
03	74682	26099	62686
04	15066	63468	12176
05	43309	69955	33319
06	74021	06418	61151
07	21464	33307	94015
08	45230	76622	74475
09	61180	00719	02573
10	98684	13136	88022
11	32336	55494	41417
12	68455	13933	32331
13	61017	30439	84178
14	86945	23972	78664
15	75679	54375	04219
16	20726	53392	39376
17	55717	70230	42594
18	58270	45287	20911
19	84518	62870	98214
20	81275	89865	77776
21	40591	67563	84947
22	34229	86481	53007
23	94897	11530	83455
24	93749	38986	42289
25	36895	46465	39832
26	91348	05054	98452
27	55369	88318	83368
28	65412	83192	74802
29	16033	59063	72705
30	47083	41745	84551
31	83603	25219	55112
32	73100	16315	96937
33	20229	89729	03264
34	01459	48858	79568
35	11420	27385	39119
36	48054	42556	69538

	A	B	C
01	88930	70684	86770
02	07502	00107	32262
03	37502	92520	27781
04	24124	20735	02885
05	88407	84476	93389
06	40930	11021	56368
07	86955	81035	73308
08	08865	54287	39807
09	05086	34039	20876
10	31184	18523	53673
11	62422	13487	42396
12	30571	42203	58053
13	49509	17409	61898
14	87976	84175	30907
15	79207	21667	36214
16	53879	90442	14257
17	28699	34124	48089
18	45749	03918	56614
19	29598	58937	91586
20	02717	96572	75120
21	26163	68513	82033
22	56181	16771	71471
23	67026	67017	64218
24	49551	92030	62925
25	45944	18478	28737
26	81906	31778	81344
27	28229	62001	28422
28	41519	28199	50066
29	56129	64627	99491
30	68401	55095	78413
31	98997	20509	75010
32	04779	97190	91164
33	38457	96056	31515
34	73700	93050	53601
35	89370	28858	29962
36	58905	32042	03995

	A	B	C
01	58933	22558	55649
02	84159	26834	58091
03	54860	27560	57695
04	76260	75156	69835
05	47346	71956	83884
06	19174	46417	58828
07	89651	19456	64310
08	85011	16115	36017
09	71419	75705	95841
10	85420	75442	06943
11	18209	72985	00874
12	59295	29292	37594
13	06039	88755	21335
14	90038	63498	96463
15	56710	84455	96414
16	98242	36632	98893
17	63209	79136	19763
18	75536	42958	46375
19	08843	94368	50007
20	19524	85549	17947
21	83088	57097	59273
22	58539	08467	22937
23	95211	99410	96785
24	63963	64831	09748
25	05630	62962	60624
26	54969	95588	70088
27	37469	99541	66877
28	17184	96424	67848
29	43815	27392	78802
30	11185	42203	16023
31	71207	99599	36808
32	39022	45742	48234
33	18383	89488	34999
34	49623	62370	77307
35	31506	11483	66973
36	65003	01494	32730

	A	B	C
01	65518	57120	74912
02	10734	58155	23242
03	11706	39782	65891
04	35187	59990	94993
05	96600	70026	47636
06	75490	89998	57555
07	92591	01958	73149
08	53126	52961	46237
09	16694	28679	63020
10	73545	37573	36699
11	23352	40260	01753
12	84892	81857	69398
13	14966	78234	14505
14	70510	84457	57236
15	43819	39980	60404
16	61455	85418	20890
17	09133	76027	62168
18	50175	23182	06263
19	59289	15262	07333
20	30442	69476	27403
21	52272	03309	19453
22	87637	97569	96179
23	33431	52908	23430
24	64771	28908	60838
25	00008	71112	60894
26	04169	04086	62778
27	61571	30524	94599
28	00100	04348	96504
29	32316	50199	66725
30	83940	80598	74823
31	65824	51570	98821
32	74216	86464	29152
33	61586	40895	50353
34	32322	10109	94284
35	33454	56895	56642
36	01537	85686	11584

	A	B	C
01	72406	75306	40887
02	80051	87764	91420
03	00328	69557	67446
04	09784	39660	17937
05	40343	53428	68988
06	69260	25756	82262
07	32181	37749	48020
08	22010	25161	16556
09	68122	96966	35768
10	01082	11888	42814
11	07632	76875	23374
12	11365	43009	06137
13	26565	63582	22218
14	71094	48531	55574
15	94849	29243	32555
16	40372	88709	91008
17	35265	92034	34746
18	49992	58910	80761
19	26392	73329	50697
20	49583	82926	10482
21	22139	44859	20993
22	49787	18532	05193
23	86747	38928	09120
24	81564	58717	19320
25	31071	55932	24134
26	31605	14977	25976
27	34636	21155	70740
28	47494	77533	25317
29	73144	27977	74386
30	81321	62049	81374
31	73301	47357	07452
32	60520	40409	54144
33	57424	37326	82118
34	18456	27071	27128
35	83370	90259	35747
36	01831	72162	87266

	A	B	C
01	98776	77729	65725
02	81079	01703	20633
03	51664	55769	86276
04	71208	96525	83412
05	20667	53573	94621
06	89551	58635	02755
07	50375	21260	16183
08	40658	49673	22702
09	04980	86809	05812
10	24272	14836	54786
11	23645	01187	70396
12	62903	49180	01689
13	58828	47705	66067
14	94248	74416	89698
15	49835	85539	11549
16	64873	22831	35399
17	19701	79760	83312
18	49747	69852	34439
19	45993	26050	11218
20	78835	50174	48199
21	21641	11387	38571
22	89118	98060	64359
23	55278	39373	04892
24	59704	60853	88129
25	95503	27722	56530
26	28454	00648	41945
27	31111	22590	17658
28	21205	15602	62420
29	76580	95997	31511
30	10975	50920	48990
31	33022	84203	61365
32	84157	77642	18797
33	39072	62073	20367
34	70540	61941	18926
35	19385	97808	22876
36	66280	01565	74079

	A	B	C
01	23496	26528	69953
02	72499	36312	43891
03	82056	46261	16724
04	09560	48373	26747
05	56564	22555	62430
06	69693	45671	53187
07	24822	98888	14222
08	81490	76458	84609
09	08199	43824	65602
10	74862	91151	00293
11	53856	91188	09315
12	93253	11967	73718
13	30185	99170	99244
14	03006	27712	62673
15	70481	18172	68796
16	16414	28693	29484
17	96788	00440	71058
18	62170	66802	13271
19	74244	58481	62795
20	42379	56066	44344
21	36694	26188	07285
22	20802	47003	99773
23	40712	44431	69817
24	70961	62258	62020
25	05524	91564	73554
26	26866	96562	90028
27	26885	90969	99986
28	61579	04677	01281
29	57548	41479	84117
30	56354	63782	67069
31	71975	42709	87080
32	66089	02892	20158
33	53494	04285	88238
34	18099	76880	47940
35	08634	45737	02015
36	47532	35144	17547

	A	B	C
01	52269	95514	44623
02	49441	97559	41233
03	25314	48909	71020
04	62135	64749	91261
05	92468	27981	83822
06	57394	56294	40405
07	39922	90129	45359
08	88026	52859	95593
09	82391	61987	04607
10	02498	92199	25396
11	94594	51063	08393
12	15680	60218	94609
13	68002	36596	39671
14	59884	17578	58750
15	63699	80508	64921
16	48194	31521	98023
17	89689	76335	61702
18	77695	71432	95188
19	33452	06284	83592
20	73132	23808	31341
21	81943	37744	41700
22	85010	27836	09477
23	64613	41858	28368
24	24581	25211	37315
25	05410	08294	77816
26	15988	76987	93832
27	74978	39390	72089
28	48462	28515	16021
29	51841	98548	73272
30	25997	09511	76639
31	04464	76120	01495
32	75168	75160	19708
33	47557	92160	93841
34	29486	23675	49496
35	44371	91510	18124
36	59911	19235	38722

	A	B	C
01	70850	32286	87858
02	76104	55530	29601
03	33039	44811	65826
04	84500	85057	87407
05	38241	97124	30637
06	38189	98749	26699
07	74677	25075	85438
08	48925	43449	24764
09	38389	16541	80687
10	70789	89815	43650
11	52937	48405	24841
12	89257	84320	47196
13	95368	28097	07711
14	73928	39937	25347
15	94039	40450	81334
16	11513	33251	20996
17	13149	68282	45538
18	08442	71326	93810
19	19948	47968	50389
20	48858	79402	64960
21	94758	76357	96033
22	90547	71528	32224
23	53913	72773	92336
24	15369	17033	57942
25	14381	80003	83823
26	69425	43426	54493
27	95974	00609	90812
28	58957	84361	08583
29	83176	10375	38570
30	82778	43168	37637
31	05233	46383	88479
32	03499	60649	03494
33	89599	77341	83524
34	41781	40138	99876
35	21042	76181	44424
36	11869	32369	71774

	A	B	C
01	34315	23454	99406
02	46370	65236	49790
03	38449	14175	31056
04	68989	05746	13856
05	25034	16450	89071
06	63353	79947	03466
07	77685	82911	14515
08	76273	15212	49150
09	51881	95143	96052
10	57368	60978	91198
11	89086	23226	23508
12	27817	78734	27318
13	63311	56263	93986
14	23733	22741	89056
15	00531	35238	77318
16	59467	84151	77528
17	92538	65983	72842
18	02810	68097	57994
19	87507	11152	37573
20	13371	42166	12642
21	69430	47948	74085
22	86622	20811	96384
23	50559	72721	17898
24	28508	19294	82189
25	17654	58936	33506
26	65444	53063	91927
27	70408	18840	63527
28	98576	03444	25980
29	26063	04818	00869
30	24839	11526	17675
31	87422	26383	58749
32	30021	03400	29793
33	24781	21655	94773
34	90050	94141	98253
35	65159	31244	99604
36	91295	69884	42166

	A	B	C
01	73641	74328	82787
02	33755	31227	97510
03	76718	65745	14504
04	13093	82400	89352
05	10168	93069	74778
06	63328	21740	14775
07	99829	06187	33806
08	99861	24988	50968
09	80020	82740	50767
10	58048	26924	24320
11	07954	11021	45626
12	79642	42819	94339
13	57322	44345	91581
14	84577	99466	04215
15	14218	81240	48278
16	04473	18681	23411
17	23352	78890	19480
18	85250	50831	61095
19	44324	91008	13447
20	12282	65141	30765
21	72645	56428	53862
22	86132	12852	25013
23	24926	96791	39982
24	82472	82164	99558
25	56953	18749	25726
26	08224	71671	72048
27	40737	28362	14084
28	29769	28015	01406
29	64536	95249	15975
30	05979	32333	89867
31	05076	14220	26474
32	13482	64772	91100
33	45302	00427	69723
34	55839	11569	68937
35	21285	95157	64961
36	87719	00143	19219

	A	B	C
01	14627	48270	26175
02	34993	89045	40501
03	60528	29855	44458
04	90306	61740	81906
05	37187	04002	94591
06	82108	36985	24438
07	16364	98545	58775
08	05290	36109	92722
09	66278	72047	39417
10	67317	30885	79716
11	62141	68569	02105
12	24464	78833	32294
13	79911	21896	35478
14	60636	54783	01643
15	15698	89298	19486
16	42698	40386	77219
17	99240	16458	56464
18	33729	85571	15792
19	86671	07699	86048
20	66963	91135	31343
21	05779	25069	32213
22	17962	88680	52082
23	52578	40742	29911
24	52024	48855	24748
25	14237	76958	29814
26	71385	77224	79645
27	78600	55187	11178
28	71563	67828	74005
29	37502	89494	40116
30	94176	11967	04026
31	34474	00602	65093
32	56674	44025	61530
33	83355	62460	45266
34	04432	87567	21839
35	02276	93225	77169
36	12930	05183	18564

	A	B	C
01	01439	17286	52242
02	86432	11302	02011
03	06451	26875	48428
04	08034	12406	51459
05	91100	76856	78111
06	24128	77821	02544
07	83284	75855	65128
08	23240	14438	42989
09	11332	47783	55828
10	93355	34648	45820
11	98326	87071	32883
12	07618	91519	99312
13	56799	23245	53317
14	39071	58427	81988
15	59071	34219	58944
16	85713	60197	48214
17	98718	38410	01713
18	29802	50460	05778
19	02654	68618	78872
20	90918	39618	97441
21	33315	76817	37684
22	92995	39636	20251
23	23783	84009	46315
24	10416	19525	86787
25	86100	73109	56334
26	81774	16366	29886
27	53935	21349	85620
28	10248	77282	75816
29	81140	93116	93200
30	09563	80343	06172
31	32416	02830	77383
32	85603	02496	71852
33	69439	96901	87518
34	01787	99960	83069
35	56971	05992	63170
36	24522	02905	22076

	A	B	C
01	47822	36959	84103
02	07416	04683	19090
03	65101	71612	94129
04	48103	97777	43780
05	03664	39247	47584
06	95386	40000	97732
07	90155	54593	23155
08	43850	98239	45326
09	49657	27155	82728
10	86749	53183	64767
11	20060	77797	27531
12	54262	10132	89562
13	69219	03161	21801
14	76959	16383	45943
15	12469	56231	91403
16	09158	97690	80921
17	22038	52264	69946
18	77349	16189	46877
19	65471	46178	80025
20	72694	89227	75549
21	30924	42384	96999
22	67063	12654	48840
23	28071	98771	92455
24	43396	37367	48638
25	57711	17999	54921
26	69394	79788	01762
27	97312	75861	21885
28	75206	80436	75379
29	94465	03309	54578
30	99003	00908	95377
31	96605	22607	70286
32	98242	95244	13244
33	53354	96733	89642
34	75953	76926	98604
35	18403	12812	87169
36	74932	86466	38178

	A	B	C
01	26116	06412	33003
02	72596	70807	02836
03	05964	20563	64877
04	64456	32388	29248
05	27663	45482	05929
06	56353	29301	49748
07	30937	10162	40764
08	74454	31249	95693
09	86541	28049	47183
10	93958	06316	03509
11	72959	11854	28961
12	96755	02453	80664
13	12453	37891	66905
14	69347	14266	15615
15	54473	56435	87149
16	08909	03495	88427
17	64592	98898	00590
18	64894	03290	84496
19	58658	14738	33280
20	54327	29421	10600
21	21217	25912	62548
22	92046	90778	78048
23	11268	56546	90246
24	64385	48032	86278
25	74367	63103	09834
26	10628	98274	34495
27	77609	94352	12670
28	69454	62375	62263
29	31094	54702	34282
30	64648	95060	73857
31	73409	05236	89062
32	82652	80659	45830
33	45674	70915	29208
34	99288	54637	35493
35	14173	11833	04026
36	03238	25047	98521

	A	B	C
01	69503	12082	85745
02	38208	87656	17611
03	62514	24200	37135
04	66085	81062	79862
05	54574	87267	83151
06	02456	37014	27577
07	64726	90881	61289
08	63940	90886	56127
09	05706	99995	96533
10	44972	78777	32468
11	89975	39860	64223
12	90058	54784	77888
13	78900	71723	41735
14	46974	27261	25177
15	09715	47544	39258
16	58072	28267	63327
17	54557	33070	77111
18	51759	76700	88461
19	33353	03444	39464
20	86610	74777	61157
21	33014	51879	28706
22	86504	17519	62779
23	70195	70411	28205
24	39446	60107	18617
25	94173	47432	17858
26	48291	06980	92950
27	11867	14252	16725
28	42168	21810	89333
29	47220	44857	59855
30	94411	31771	19945
31	42184	15763	74197
32	50880	61799	99437
33	87764	09193	58870
34	31887	36202	50011
35	26311	59096	52832
36	77223	56479	00054

	A	B	C
01	51198	85428	41604
02	02779	87845	42525
03	31678	58150	35960
04	89707	09224	22908
05	36986	43707	21530
06	94478	04097	76156
07	64575	10793	29190
08	82717	22968	87311
09	14133	59723	94499
10	09668	56883	19590
11	82559	47451	47161
12	60014	42909	24232
13	68487	08122	59897
14	14608	00119	22138
15	81308	56079	22582
16	61409	66694	73011
17	46529	08579	71392
18	53391	28910	11775
19	47919	44112	35268
20	47094	35217	96311
21	12744	98155	45886
22	82067	22270	37330
23	96454	98534	19069
24	68666	64600	70077
25	78941	61108	92975
26	50106	99499	74984
27	81766	23285	24422
28	83086	88161	45834
29	50040	05297	32779
30	48631	56685	77749
31	09960	05338	30761
32	17637	86612	39427
33	05902	10581	30516
34	53042	01990	58484
35	20424	47769	84223
36	40015	91504	68535

	A	B	C
01	09212	55675	03831
02	07387	78745	96123
03	06983	83409	83001
04	81781	28936	19102
05	59483	46171	04385
06	77249	80127	48152
07	00496	82340	56515
08	16088	16949	25617
09	07283	64828	03029
10	99156	68008	54657
11	81622	67311	46457
12	21465	99369	19751
13	19556	17490	74719
14	69496	15138	55213
15	28912	49634	98939
16	77860	28429	88390
17	74338	74735	68766
18	29247	75493	57844
19	27614	14226	73999
20	35532	26184	57190
21	29327	41530	18713
22	18368	37834	35860
23	32716	00421	05472
24	66008	17885	18007
25	32404	94074	28683
26	63069	18207	76877
27	86524	26150	62917
28	56567	86045	67587
29	02970	11889	90431
30	49699	32404	50393
31	43084	06468	76629
32	97944	38503	09740
33	05967	37369	74566
34	65418	49203	61416
35	48662	34438	36165
36	47180	08986	71921

	A	B	C
01	73577	23904	77216
02	15987	60513	63840
03	80648	20897	61121
04	71244	29817	47714
05	91618	70757	44666
06	30865	60928	45397
07	04055	65381	42854
08	10152	69963	73731
09	97042	76166	25205
10	91448	34181	36026
11	34900	52921	89444
12	53388	92292	73290
13	07303	01385	43945
14	67924	25866	42715
15	71536	00385	33750
16	14352	08225	52743
17	23201	87887	92077
18	85225	40243	27152
19	61461	70167	69695
20	78711	62716	81954
21	90065	79891	82557
22	54466	43504	97950
23	75493	39556	07636
24	98088	59977	96004
25	39125	86238	16385
26	85786	55747	31347
27	10912	43793	24735
28	68264	37227	46268
29	63111	38341	59087
30	39917	99473	23742
31	47143	67931	82633
32	49617	16781	05307
33	48553	88237	43032
34	40125	60961	17361
35	04375	69639	57647
36	16960	85578	23367

	A	B	C
01	90060	72465	16576
02	33565	49072	16712
03	86383	24902	25141
04	47708	51433	69400
05	39697	18524	74389
06	79697	54400	10578
07	59418	64964	29774
08	04823	09778	72898
09	77505	65731	59975
10	49460	64361	82095
11	55067	69064	93185
12	29773	12721	01550
13	75630	74670	47588
14	24552	29563	33036
15	46939	47319	68651
16	14748	38793	53141
17	80528	18456	33706
18	69354	45489	44763
19	16306	21410	89097
20	63562	80805	98178
21	41528	74247	48457
22	29994	61161	33327
23	71581	09157	04918
24	41791	00534	02882
25	56165	56903	19797
26	44765	81320	95035
27	11272	78565	77593
28	29567	23396	40298
29	84987	38633	74017
30	12082	41971	57235
31	47671	53644	64264
32	30602	88474	06031
33	38407	07594	95482
34	63562	32403	01386
35	40762	72110	15814
36	95860	98963	24517

	A	B	C
01	08985	00339	40851
02	44950	29575	03055
03	42544	44362	26016
04	32333	75283	99631
05	74532	68340	88757
06	55659	19237	86829
07	37830	80684	01566
08	17089	83856	01109
09	14423	34306	05444
10	02651	31837	29728
11	97153	01191	56687
12	73316	30316	42618
13	01740	85052	73791
14	02235	26789	32513
15	03933	99309	37244
16	73712	47998	10937
17	85953	67864	83610
18	96385	80121	64624
19	46167	34919	03033
20	97722	56108	57323
21	81561	72128	96150
22	65752	25245	64259
23	70932	76700	58799
24	76118	22560	83553
25	98342	10075	16970
26	59042	41658	51967
27	14558	19862	67461
28	05600	58208	20262
29	03642	12334	45616
30	51421	03500	10278
31	72205	80639	97267
32	31080	81133	80851
33	17774	72333	53307
34	87258	53329	80577
35	94545	04155	08582
36	43987	14601	43215

	A	B	C
01	57187	10775	01538
02	95523	33928	16075
03	51235	72542	58378
04	74404	63681	99340
05	57484	54302	47776
06	43052	16989	98638
07	78859	05144	82858
08	60234	99288	09350
09	35661	97548	37460
10	40195	99049	88610
11	71860	72592	74768
12	36646	23205	79516
13	63886	14838	17522
14	52125	96143	86415
15	20234	29607	63024
16	85434	10221	02396
17	93035	02442	68146
18	95822	81850	95480
19	46478	63694	93246
20	60932	73630	16375
21	95076	28066	57918
22	17200	17698	12233
23	37572	88880	04994
24	52516	09881	72968
25	70886	05047	68160
26	75484	11567	59570
27	14181	92466	05618
28	70686	91851	73888
29	64053	22998	60941
30	35736	79265	33785
31	50715	41197	35322
32	21254	40034	41554
33	70467	54706	59877
34	87965	58391	91928
35	97986	23635	76568
36	16213	97930	45871

	A	B	C
01	45874	90710	23741
02	97050	15386	27943
03	80688	92496	22053
04	52653	90439	54417
05	73138	18202	69918
06	72246	40250	29466
07	47884	58873	71221
08	29561	83683	98725
09	98648	64734	51251
10	26883	76750	75366
11	81620	63692	18959
12	77633	35415	81826
13	13222	06740	62005
14	97462	25279	36745
15	42787	60984	71216
16	98071	75429	65121
17	19487	12394	71275
18	55446	09237	40750
19	96541	33557	67340
20	41845	39538	97710
21	34460	46988	50012
22	40667	38142	81034
23	89003	03780	89028
24	74633	22649	15012
25	97868	63876	47392
26	74089	03965	41943
27	21248	23624	88854
28	61712	71015	13592
29	45406	39952	71304
30	31405	43135	33646
31	33582	35788	20011
32	90483	44251	57747
33	95268	31564	76070
34	39723	80998	67121
35	48643	56885	60931
36	95635	40533	05754

	A	B	C
01	96451	03997	22923
02	67433	50590	52955
03	09411	26681	26167
04	50300	27071	19025
05	33058	54424	36906
06	32496	34186	61814
07	01572	90708	81484
08	72764	64572	28332
09	91140	78345	51066
10	95972	07619	25602
11	92751	91073	83715
12	03263	25780	95151
13	85470	27206	18553
14	76695	44392	31112
15	16820	77017	22608
16	34869	91294	42220
17	40001	36326	27642
18	74506	65919	15474
19	34389	78062	63615
20	52961	06854	83104
21	55876	10809	74927
22	61528	79525	99555
23	50985	59174	76332
24	39611	15957	14778
25	77110	97713	84869
26	78089	70317	08872
27	66286	45683	31182
28	92994	18926	54875
29	14278	91801	45744
30	59875	20000	97677
31	32487	65401	15342
32	46706	33308	94595
33	65305	50481	74515
34	34999	65082	44599
35	02122	17761	69339
36	81285	95170	43871

	A	B	C
01	22563	62771	13554
02	96894	30833	81589
03	05616	29027	89977
04	92730	18507	78531
05	69211	10782	84450
06	23873	25366	65896
07	43899	33841	56001
08	30914	61288	68001
09	31975	61295	95078
10	93778	95379	01612
11	35262	50672	88000
12	24725	73739	06369
13	20739	53176	71044
14	09301	18959	35120
15	67284	27153	16001
16	77976	58431	82939
17	29558	77456	60435
18	48568	76551	26200
19	62025	98900	40711
20	23926	16825	16914
21	54510	49208	86350
22	35887	16677	92748
23	32312	94112	03724
24	42538	03259	51748
25	41523	59720	26292
26	61507	84768	80493
27	31032	65632	37469
28	46511	68379	89859
29	54969	13444	82464
30	35287	03049	91182
31	63977	83632	31122
32	68801	50542	78726
33	49240	24893	07338
34	75359	08211	77310
35	85939	88373	83330
36	41090	43458	80594

	A	B	C
01	70868	57000	38968
02	49301	60760	66217
03	04960	12542	73624
04	39897	96225	02406
05	39036	43131	72621
06	03228	50946	40221
07	87911	92081	10526
08	63961	92353	04224
09	25284	61034	87351
10	76280	13245	83917
11	92627	96330	09661
12	26215	20368	21793
13	57133	16127	14112
14	62323	67552	76142
15	95562	15504	46158
16	98857	98363	16876
17	07774	21221	88580
18	61907	58502	33367
19	17448	25461	91442
20	99591	85733	19381
21	97481	22920	02402
22	17143	35916	20721
23	57773	99973	34408
24	59593	08554	62444
25	01030	34055	92198
26	60858	69447	38491
27	44788	38803	62271
28	66299	52336	57802
29	66169	43215	04511
30	36809	42013	44465
31	73382	66107	42209
32	22967	39116	52298
33	74030	89751	45279
34	70384	93735	49462
35	15275	35194	95470
36	79437	76134	52544

	A	B	C
01	49541	73705	59772
02	35066	94632	44796
03	94990	66839	31824
04	57324	93729	04427
05	16495	76668	01932
06	17097	37232	17305
07	72142	02871	73520
08	09641	09230	22011
09	54981	13954	66086
10	70610	03766	66364
11	09393	83885	80674
12	74913	85005	42086
13	91101	08888	45301
14	48466	75876	68314
15	65531	80862	71680
16	85078	82438	77013
17	51906	39196	74122
18	48702	22101	73517
19	15677	15292	04289
20	15974	08643	98649
21	63642	65466	93461
22	00571	38885	66656
23	47294	32924	49299
24	06876	71691	00252
25	79503	07295	57488
26	56152	74242	21705
27	88464	34945	19824
28	95157	25237	96354
29	77539	62409	09317
30	56555	96126	34081
31	29107	03726	70910
32	48575	02287	25254
33	74467	30929	78338
34	28040	61720	77138
35	47963	04362	93443
36	94324	21235	91553

	A	B	C
01	15284	60095	84643
02	39942	43028	48178
03	13218	80436	36884
04	12616	01208	72516
05	17902	83586	80712
06	78538	83871	01436
07	45537	33829	33433
08	05836	27128	12209
09	04687	30790	76777
10	75760	64298	07061
11	92709	95584	88852
12	30439	88620	46032
13	43596	02653	00188
14	03451	07181	91134
15	46710	29486	78406
16	24286	98854	85141
17	21806	97858	96153
18	15969	06592	93778
19	32691	50911	14026
20	13150	14842	21248
21	69153	89536	03260
22	51613	23739	20986
23	06868	03849	30746
24	16422	79000	46041
25	97327	23119	08663
26	57454	16439	83222
27	80971	95343	28381
28	92620	90033	43782
29	47507	43690	54816
30	87609	88015	58536
31	56381	89395	95135
32	12853	48583	57684
33	17067	03937	35526
34	11826	16119	85160
35	06243	77326	86132
36	86973	74286	49001

	A	B	C
01	22751	13033	29113
02	25460	41303	40907
03	75850	71082	88416
04	63437	91697	31524
05	79508	99656	18424
06	87215	47404	33327
07	97057	79354	15481
08	90219	80710	52509
09	43440	23586	72806
10	53712	16374	15706
11	13742	89019	80616
12	81740	81713	90973
13	73230	03729	33784
14	76778	44067	00785
15	15345	74647	91167
16	97031	97837	69418
17	81777	48142	70643
18	68883	90365	41100
19	23464	57794	39993
20	30364	19733	69810
21	89553	92604	79779
22	85720	66920	38988
23	49524	17089	33469
24	90990	88515	91506
25	05291	24177	37799
26	89626	38399	28341
27	57334	50238	29785
28	54541	43993	95639
29	23246	89013	06775
30	98736	00534	18471
31	63096	28410	29094
32	17955	14799	92058
33	43914	76910	03259
34	55516	46396	29940
35	37537	63568	80109
36	54313	48145	49842

A

00	89660
01	74877
02	20045
03	89907
04	60690
05	20013
06	92845
07	75805
08	19970
09	50684
10	61319
11	97888
12	46360
13	36341
14	54374
15	27205
16	68023
17	92105
18	53477
19	43466

Notes

Notes

Notes

Notes

Notes

Notes

Notes